Creationism, Science,
and the Law:
The Arkansas Case

Creationism, Science,
and the Law:
The Arkansas Case

edited by
Marcel C. La Follette

The MIT Press
Cambridge, Massachusetts
London, England

This book was set in Times Roman
by The MIT Press Computergraphics Department and printed and bound
by Halliday Lithograph in the United States of America.

Library of Congress Cataloging in Publication Data

Main entry under title:

Creationism, science, and the law.

  Bibliography: p.
  Includes index.
  1. McLean, Bill. 2. Arkansas. State Board of Education. 3. Creationism—
Study and teaching—Law and legislation—Arkansas. 4. Science and law.
I. La Follette, Marcel Chotkowski.
KF228.M39C73 1983   344.73′077   82–21646
ISBN 0–262–12101–8   347.30477
ISBN 0–262–62041–3 (pbk.)

Contents

Contents

Contents

Acknowledgments

Many of the essays and documents in this volume were originally published in the Tenth Anniversary Issue (Summer 1982) of *Science, Technology, & Human Values*, a quarterly journal sponsored by the Program in Science, Technology, and Society, Massachusetts Institute of Technology, and the John F. Kennedy School of Government, Harvard University. Special thanks are due to the MIT Program for general support of the journal during the development of Issue 40 and this book.

The editor of any collective work resembles an orchestra conductor—a successful performance possible only because of the talents of composers, patrons, and musicians. This book owes more than is usually the case to its authors; they are representative of several dozen attorneys and witnesses who participated in the Arkansas case. The journal's Editorial Advisory Board, and especially William A. Thomas and Gerald Holton, gave greatly appreciated support, advice, and encouragement at critical points throughout this project. And special thanks go to Melinda Thomas, Lisa Buchholz, and David Grant, without whom the effort would have been, if not impossible, then surely less fulfilling and affirming of the best in the human spirit.

Contributors

Ann Rebecca Bleefeld has taught biology and was working as a secretary at Skadden, Arps, Slate, Meagher & Flom during the Arkansas case (*McLean v. Arkansas Board of Education*). She served as a consultant to the science team for the case.

Stephen G. Brush is Professor of History at the University of Maryland, Institute for Physical Science and Technology.

Joel Cracraft, Professor of Anatomy at the University of Illinois Medical Center, was science advisor to the American Civil Liberties Union for the *McLean* case.

Gary Crawford, an Associate with Skadden, Arps, Slate, Meagher & Flom, was one of the trial attorneys in the *McLean* case.

Langdon Gilkey, Shailer Matthews Professor of Theology at the Divinity School of the University of Chicago, testified for the plaintiffs in *McLean*.

Mark E. Herlihy, an Associate with Skadden, Arps, Slate, Meagher & Flom, participated in *McLean*.

Eric Holtzman, Professor and Chairman of the Department of Biological Sciences at Columbia University, advised the American Civil Liberties Union during development of the *McLean* case.

Peggy L. Kerr, a partner in Skadden, Arps, Slate, Meagher & Flom, was in charge of the firm's work in *McLean*.

David Klasfeld, an Associate with Skadden, Arps, Slate, Meagher & Flom, was one of the trial attorneys in *McLean*.

Marcel C. La Follette is Editor of the journal *Science, Technology, & Human Values*, and Research Associate, Program in Science, Technology, and Society, Massachusetts Institute of Technology.

Larry Laudan is Visiting Research Professor at the Center for the Study of Science in Society, Virginia Polytechnic Institute and State University, and Professor of the History and Philosophy of Science at the University of Pittsburgh.

Dorothy Nelkin, Professor in the Cornell University Program on Science, Technology, and Society, testified for the plaintiffs in *McLean.*

Michael Ruse, Professor of History and Philosophy at the University of Guelph, testified for the plaintiffs in *McLean.*

Susan Sturm, Staff Counsel at the American Civil Liberties Union, participated as an attorney in *McLean.*

Introduction   Marcel C. La Follette

On 19 March 1981 Arkansas Governor Frank White signed into law
Act 590, "The Balanced Treatment for Creation-Science and Evolution-
Science Act." The legislation had been debated in the Arkansas Senate
for only five minutes, no hearings were held in either house, no scientists
testified for or against passage, and Governor White has since stated
that he did not read the act before signing it.[1]

Act 590 was essentially identical to a "model bill" drafted by Paul
Ellwanger, the founder of Citizens for Fairness in Education, an organization based in South Carolina. Ellwanger's bill, and its embodiment
in Act 590, represent the political expression of a social movement
based on the belief that all life in the universe was created by a supernatural being (God), out of nothing, within a seven-day span, 6,000 to
20,000 years ago, as described in the Bible in the first chapters of
Genesis.

Believing Act 590 to be a disguised attempt to establish religion in
the public schools, in violation of the First Amendment of the U.S.
Constitution, a group of twenty-three Arkansas citizens and organizations brought suit on 27 May 1981 in U.S. District Court.* Trial in

---

*The individual plaintiffs included: the resident Arkansas Bishops of the United
Methodist, Episcopal, Roman Catholic, and African Methodist Episcopal
Churches; the principal official of the Presbyterian Churches in Arkansas; other
United Methodist, Southern Baptist, and Presbyterian clergy; and several persons
who sued as parents and next friends of minor children attending Arkansas
public schools. Among the organizational plaintiffs were the American Jewish
Congress, the Union of American Hebrew Congregations, the American Jewish
Committee, the Arkansas Education Association, the National Association of
Biology Teachers, and the National Coalition for Public Education and Religious
Liberty.
The defendants included the Arkansas Board of Education and its members,

Rev. Bill McLean et al. v. Arkansas Board of Education[2] commenced 7 December 1981 and lasted two weeks. The court's opinion, handed down on 5 January 1982 by Judge William R. Overton, held the act to be unconstitutional on the grounds that (1) it had been passed with the specific purpose of advancing religion, (2) it had as a major effect "the advancement of particular religious beliefs," and (3) it created for the state of Arkansas an "excessive and prohibited entanglement with religion." Act 590 was thus a law quickly passed and quickly overturned.

The implications of the Arkansas case extend, however, beyond the borders of one Southern state and beyond the walls of public schools or Federal courtrooms. Litigation on creation-science bills continued throughout 1982. The evangelical spirit that inspires the creationists and the social forces that support their political efforts will not dissolve because of one adverse court decision. In fact, many of the major actors in *McLean v. Arkansas* were neither residents of Arkansas nor engaged in public-school teaching. Arkansas, "The Land of Opportunity," was simply a platform for one act of an ongoing struggle between "fundamentalists" and "free-thinkers," just as the Scopes trial in a small Tennessee town had been fifty-six years before.

Because the Arkansas case encompassed the gamut of legal and political action (lobbying, legislation, constitutional challenge, trial, and judicial opinion), it affords an excellent framework within which to examine key points of law and social debate. Moreover, because the case dealt with a recurrent conflict between science and religion, it tells us something about the general public's uneasy relations with science, about the problem of defining what is or is not science, and about the question of who should properly decide that definition within an educational setting. The present book brings together the pertinent legal documents in the case and discussions by some of its participants and close observers. The essays have been commissioned to give some flavor of the debate, to exhibit a variety of perspectives on the issues raised and perhaps to raise some new ones, and to guide readers to other, more exhaustive sources. This book is not a primer on either Creationism or evolutionary theory, nor is it an attempt to construct an artificial confrontation between creationists and evolutionists. The

---

the Director of the Department of Education and the State Textbooks and Instructional Materials Selecting Committee. The Pulaski County Special School District and its Directors and Superintendent were voluntarily dismissed by the plaintiffs at the pre-trial conference held 1 October 1981. The State of Arkansas was dismissed as a defendant because of its immunity from suit under the Eleventh Amendment (*Hans v. Louisiana*).

intention is simply to examine some of the social and philosophical implications of *McLean v. Arkansas.*

Creationism and Evangelism

Present-day creationists trace their roots to the evangelical Protestant reaction to the theories of Charles Darwin. In the nineteenth century, in England, opposition to evolution was a test of one's faith in the flawlessness of the literal interpretation of the Bible.[3] In the 1920s, in America, fundamentalist protests against the teaching of evolution drew strength from a questioning by many Americans of contemporary social values and by an increasing anxiety over the rapid social changes that often accompanied new developments in science and technology. To the fundamentalists, the teaching of evolutionary theory seemed connected to the decay of moral values—a particularly forceful accusation when popular sentiment suggested that the country was "going to ruin." These feelings boiled over in the 1920s in Dayton, Tennessee, when a high-school biology teacher, John T. Scopes, was tried and convicted for teaching the theory of evolution.[4] Dorothy Nelkin's essay in this volume attributes the recent resurgence of public controversy over the teaching of evolution to similar perceptions of eroding traditional values and growing secularism.

Reverend Tim LaHaye, founder of the Institute for Creation Research, believes that "the chaos of the '60s is a result of teaching evolution," and warns that "the elite-evolutionist-humanist is not going to be able to control education in America forever."[5] A fundamentalist writing to the *Arkansas Democrat* during the *McLean* trial complained that "your Darwinism Theory is a nightmare of horrors!"[6] But such complaints are, according to biologist Philip Kitcher, part of a strategy: "by depicting evolution as opposed . . . to a large number of institutions that are cherished by a large number of people," the creationists are attempting to "make it 'easier' for their audience to choose sides" in the "genuine" conflict between evolutionary theory and a literal reading of the Book of Genesis.[7]

Although creationists identify evolution as a Satanic source of social ills, their tactics for placing "creation-science" in the public schools take full advantage of the unquestioned authority of science in modern society. Langdon Gilkey points out in his essay that "fundamentalist and cultic forms of religion have grown in our lifetime *because of*, not in spite of, the dilemmas of a technological society." It is this willingness to *use* science to advance their cause that sets modern creationists apart from their predecessors. In a society that values science, the appearance of conflict between science and religion could, the creationists seem to believe, seriously disturb religious faith; so, as Gary Crawford points out in this volume, they have attempted to reconcile their beliefs with modern science by using empirical data to prove events described in the Bible. The data are then realigned within a framework of revelation. For scientific creationists, the Bible is not a myth; all biblical "facts" can be proven by "scientific" means.

The creationist organizations, such as the Institute for Creation Research, promote in their literature the formal scientific credentials of their staff members. Such efforts, and the coining of the terms "creation-science" and "scientific creationism," represent attempts to gain public credibility, a strategy that relies on the relative scientific illiteracy of most Americans. One prominent chemist has stated that when first confronted by the widespread acceptance of creationists' campaigns he couldn't understand the "gullibility" of people: "It took me a while to understand that the average American is not equipped to combat this sort of thing."[8] How can an ordinary person weigh the scientific quality of the work of two researchers—one of whom espouses creation-science, the other evolution—when both hold degrees from well-known universities and both use identical scientific terms to discuss the fossil record of animals?

Proving creation-science theories is a critical part of the creationists' evangelical mission, their way of witnessing for Christ. The Bible Science Association, for example, states that "we have discovered that when we witness to people on the basis of science and show them the need of Christ and His spirit in the sciences, we have one of the most powerful methods of witnessing the Church today."[9] This is representative of the new evangelism of American fundamentalism, which involves sophisticated political sense and the use of the judicial system and the mass media as well as of conventional social-evangelical methods to achieve religious goals.

Because creationists believe it is impossible to devise a legitimate means of harmonizing the Biblical account of creation with evolution, they see only two possible positions with respect to the origins of life on earth and argue that, in the interests of "balanced treatment" and fair play, the public schools must include a "two-model" approach to the teaching of origins. Creation-science is one model of origins; evolution, another. Either the universe is the work of a Creator, or it is not.[10] When both approaches cannot be discussed equally, then neither should be discussed.

In Act 590 and other pro-creationist legislation, "creation-science" is defined as the scientific evidence (and the scientific inferences from that evidence) indicating (1) "the sudden creation of the universe, energy, and light from nothing"; (2) "the insufficiency of mutation and natural selection in bringing about development of living kinds from a single organism"; (3) that changes occur only within fixed limits of the original kinds of plants and animals; (4) separate ancestry for man and apes; (5) explanation of the earth's geology by catastrophism, including the occurrence of a world-wide flood; and (6) the "relatively recent" inception of the earth and of biological "kinds." The legislated definition of "evolution-science" is more or less a mirror image; (1) the emergence of life by naturalistic processes and disordered matter; (2) the sufficiency of mutation and natural selection in bringing about the development of present living kinds; (3) emergence by mutation and natural selection; (4) emergence of man from a common ancestor with apes; (5) explanation of the earth's geology in evolutionary sequence through uniformitarianism; and (6) an age of the earth set at several billion years.

The plaintiffs in *McLean v. Arkansas* argued that Act 590's two-model approach was a deceptive attempt to introduce religious beliefs into the public schools. In their brief, the plaintiffs declared that they were not seeking to censor creation-science as a religion but, "out of respect for science and out of reverence for religion," to maintain the separation of Church and State. Mandatory inclusion of creation-science, they believed, would force what they believed to be a false choice between science and religion, and ultimately between science and God.

Legal Precedents

The plaintiffs' challenge invoked a constitutional question: Was Act 590 in violation of the Establishment Clause of the First Amendment of the U.S. Constitution?

Two precedent cases, *Epperson v. Arkansas* and *Daniel v. Waters*, pertain directly to the trial strategy and opinion in *McLean*. The earlier case, *Epperson v. Arkansas*,[11] centered around a challenge to the constitutionality of a 1928 Arkansas antievolution statute prohibiting state-supported schools or universities from teaching "the theory or doctrine that mankind ascended or descended from a lower order of animals." Susan Epperson, who was a tenth-grade biology teacher in the Little

Rock public school system, brought the case in 1965 when faced with a dilemma of using a new text, adopted by the County School Board for instructional use, which "contained chapters setting forth 'the theory about the origin . . . of man from a lower form of animal.' " According to the 1928 Arkansas statute still in force, teaching evolution theory would have been a criminal offense and could have subjected Epperson to dismissal; hence the constitutional challenge. In 1968, the U.S. Supreme Court held that the statute violated the First Amendment because the sole reason for its passage was that a particular religious group considered the theory of evolution to conflict with the account of the origin of man as set forth in Genesis:

Government in our democracy, state and national, must be neutral in matters of religious theory, doctrine, and practice. It may not be hostile to any religion, or to the advocacy of no religion, and it may not aid, foster, or even promote one religion or religious theory against another, or even against the militant opposite. The First Amendment mandates government neutrality between religion and religion, and [between] religion and nonreligion.[12]

The 1928 statute had not sought to remove all discussion of the origin of man from the school curricula, but it had been "confined to an attempt to blot out a particular theory because of its supposed conflict with the Biblical account, literally read."

In the other precedent case, *Daniel v. Waters*,[13] challenge was made to a 1973 Tennessee law that prohibited any type of book expressing an opinion "about the origins of creation of man in his world" from being used as a public school textbook "unless it specifically stated that it is a theory . . . and not represented to be scientific fact." The statute also required that "commensurate attention . . . and an equal amount of emphasis" be given to the account of creation as set forth in Genesis, although the Biblical account would carry no disclaimer regarding its status as theory or fact. The U.S. Court of Appeals, Sixth Circuit, held that the result of the Tennessee legislation was a "clearly defined preferential position for the Biblical version of creation as opposed to any account of the development of man based on scientific research and reasoning." In assigning such preferential position, the law violated the First Amendment.

As defined by the U.S. Supreme Court, the Establishment Clause in the First Amendment ("Congress shall make no law respecting an establishment of religion, or prohibiting the free exercise thereof") means at least this: "neither a state nor the federal government can set up a

Church," can pass laws which "aid one religion over another," or which aid all religions, nor can a state "influence a person to go or to remain away from church against his will, to force him to confess a belief or disbelief in any religion," nor can people be punished for "entertaining or professing religious beliefs or disbeliefs."[14] Thus, as Judge Overton wrote in the *McLean* Opinion, the Establishment Clause "enshrines two central values"—voluntarism (that is, a person cannot be forced to believe or not to believe, to do or not to do, something vis à vis a particular religion) and pluralism (that is, the belief that there are many religious beliefs and that all should be tolerated under the Constitution). The Supreme Court, in the development of law on this point, has generally formulated a test for whether a statute meets the requirements of the Establishment Clause. As articulated in *Lemon v. Kurtzman*,[15] (1) a statute must have "a secular legislative purpose," (2) "its principal or primary effect must be one that neither advances nor inhibits religion," and (3) "the statute must not foster 'an excessive government entanglement with religion.' "

But how do such legal formulations relate to philosophical definitions of science? The answer lies in creationists' choice of social ministry. In order to place their religious values within the public school system, the creationists argue that creation-science is a science. Therefore, to prove the Arkansas statute unconstitutional, the plaintiffs did not need to prove that creation-science is a religion, only that it *advances* religion. As Gary Crawford points out, "the determinative Constitutional question is not into which philosophical classification creation-science fits, but whether the function . . . of teaching creation-science in the public schools is to promote belief in sectarian doctrine and thereby render the schools partisan to particular religions."

At first, the attorneys for the plaintiffs felt that it would be "easy enough to define science and then to demonstrate that creation-science is not science."[16] Point-by-point refutation of creation-science was rejected as a strategy, however, because it might have given the impression of a contest between positions of "comparable weight" and hence shore up the argument of balanced treatment for two models. Ultimately the legal strategy chosen was to focus on creation-science, not on evolution. Expert witnesses were chosen to testify around three segments of the case: science, religion, and education. The *religion* proof showed a legislative history to the bill which exhibited a religious purpose. Moreover, as Langdon Gilkey argued in his testimony, the concept of a creator or supernatural creator inherent in creation-science does not differ from the concept of God. In the *science* case, the plaintiffs testified to the scientific evidence against the creation scientists' claims. Joel Cracraft outlines in his essay six of these major assertions. Simple use of scientific techniques and terminologies, the plaintiffs also showed, does not make something science. Philosophers such as Michael Ruse were called to testify that science is rooted in actual laws, not in supernatural persons, events, or processes. The plaintiffs' brief argues:

> . . . a system of belief that has as its center the interruption, suspension, or non-existence of natural laws and places in lieu thereof the intervention of an omnipotent Creator (God) is not science. Even if some of its minor premises look, smell, taste, and feel, and sound scientific, its major premise, God, is not subject to testing and to disproof and is accordingly not scientific. Such a system of belief, not being science, is religion.

The *education* case for the plaintiffs attempted to show that the secular educational value of creation-science was nil. Without a religious framework, without God, creation-science is simply an assortment of facts and assertions; with God, it is religious and therefore constitutionally forbidden to be taught in the public schools. The logical alternative to teaching creation-science under Act 590 would be to eradicate the study of evolution from public school curricula in Arkansas.

The plaintiffs' strategy centered not only on the philosophical structure of science and whether a science is always based on natural laws, but also on the social structure of science. The plaintiffs attempted to prove that creationists appeal not to natural but to supernatural explanations of physical events, and that they insulate their viewpoints from the rigors of empirical verification. Time after time, creationist witnesses admitted in testimony that they had not actually submitted their papers to established peer-reviewed scientific journals. Philosophers of science testified that creation-science started with an immutable a priori assumption that the Bible is inerrant and is historically and scientifically true, a fact that new members of many creationist organizations must affirm by oath.[17]

In defense of the statute, the Arkansas Attorney General argued that both *Epperson v. Arkansas* and *Daniel v. Waters* provided proof of constitutionality. The decision in *Epperson* turned upon the finding that a scientific theory (evolution) was being censored for religious reasons. The state argued that in this case the "proverbial shoe was on the other foot" because the *McLean* plaintiffs were seeking to "censor a theory [creation-science] which, while perhaps controversial, is well-supported by competent scientific evidence." Under this interpretation, the decision in *Epperson* did not preclude the inclusion of creation-science in the curriculum, but only addressed matters such as the exclusion of the theory of evolution for religious reasons. Because Act 590 mandated "a neutral, two-model approach" for reasons relating to education and science but not to religion, the *Epperson* decision could be interpreted as supporting the constitutionality of Act 590. The decision in *Daniel v. Waters*, the state argued, provided similar support for constitutionality because "the key to the holding in *Daniel* was the preferential position given the Genesis account of creation"—that is, the court had not declared that *any* inclusion of the Biblical view of creation was invalid, only that "favoritism" of the Biblical view established religion. To the state, the decision in *Daniel* showed that a legislature could require diversity in the presentation of "theories of origins" whenever such theories were presented. This, according to the state, was exactly what Act 590 was intended to do; that is, Act 590 advanced both scientific inquiry *and* academic freedom because students were provided with alternatives. Academic freedom would be damaged only if a "controversial" theory (creation-science) was "squelched or censored by any self-appointed guardians of what is 'science.' "

The Opinion in the Arkansas Case

In ruling Act 590 unconstitutional, Judge Overton states that Act 590 conveys "an inescapable religiosity" and was passed with the specific purpose of advancing religion. The *McLean* decision notes that the definition of creation-science given in the legislation referred to supernatural creation by God, and that the provisions of the statute were "not merely similar to the literal interpretation of Genesis" but "identical and parallel to no other story of creation."

The opinion further states that creation-science is not science because it is not explainable by reference to natural law, is not testable, and is not falsifiable (a ruling that relied heavily on evidence from the philosophy of science). Furthermore, creation-science also fails to meet the standards of science because it uses terms that have no scientific meaning, and the work of creation-scientists is not published in accepted scientific journals where it would be subject to peer review and testing. Creationists, the opinion asserts, do not weigh data against opposing data in order to reach conclusions, as scientists do; instead, they look to the literal wording of the Bible and then attempt to find scientific support. The finding that creation-science has no scientific merit or educational value as science is legally significant when linked to the major effect of the Act: "Since creation-science is not science, the conclusion is inescapable that the *only* real effect of Act 590 is the advancement of religion."

Controlling the Definitions

The question of who should determine the content of science education underlies both the Arkansas statute and the plaintiffs' challenge. The statute, and others like it, spring out of widespread fear by parents that they are losing control over what their children learn, and many Americans' fear, irrespective of religious affiliation, that they are losing control over their own lives.[18] Fundamentalists, particularly in the southern and western United States, believe that their actions have majority support. They regard the "evolutionists" and "secular humanists" as the monopolistic minority. Moreover, creationists regard themselves not only as parents fighting for their children's minds, or as religious people "fighting for Christ," but also as a new political and patriotic force. Their campaign is reinforced by a general perception of deterioration in "basic American values." For example, in a May 1981 *Time* magazine/Yankelovich, Skelly, & White survey, 73 percent of the public polled agreed with the statement that "the Supreme Court and Congress have gone too far in keeping religious and moral values out of our laws, our schools, and our lives."[19] Sixty-two percent of the respondents in a 1981 Gallup poll felt that schools paid insufficient attention to developing moral and ethical character. Other polls show a significant decline of confidence in public schools (for example, 40 percent in 1981 from a high rating of 60 percent in 1973), and a majority feeling that local school boards, not state government, should have the greatest influence.[20]

The educational establishment maintains that school boards, the state, and local textbook selection committees, *not* the "majority" rule of parents or special interest groups, should determine the issues of concern in their own areas. In such roles the educators may indeed be experts, but they are also the vicars of society, authorized to promote the state's legitimate interest in the education of its citizens, and to select for instruction the information that best prepares students to become members of society. Americans have made the assumption that education in science and in the scientific method is necessary for all citizens. Polls confirm that a vast majority of the public (in some cases over 90 percent) favor at least two years, if not four years, of science and mathematics for students in high school, even for those who do not go on to college.[21]

The legitimate authority to instruct, though, is tempered by recognition that public schools are institutions in which the state both controls and influences participants. The principle of academic freedom has therefore been invoked to limit the state's power to interfere with education in the classroom. The Arkansas statute, the plaintiffs argued in their brief, damaged academic freedom because it not only froze scientific definitions but also required "to be taught as science a view that has not yet proven itself and gained acceptance within the scientific community."

Introduction of this argument (and Judge Overton's use of it in his opinion) has shifted the question of definitional control to scientists and philosophers of science and has claimed for scientists a special privilege as the only arbiters of the rules, orders, and definitions of science. In the legal forum of the *McLean* case, plaintiffs asserted that a specialized community (scientists, working with science educators) had a legitimate right "to determine issues bearing upon their particular competence."

Most people would agree that the content of science classes should be determined not by a consensus of "the votes of shopgirls and farmhands,"[22] but by a consensus of the best scientific opinion. Yet is it not equally risky to place legal imprimatur on the assertion that, by virtue of shared experience and training, members of the scientific community are the sole authorities on what is science? Mark Herlihy and Larry Laudan, in their essays in this volume, touch on some of the dangers inherent in interpreting the Overton opinion as granting legal status to a particular definition of science, rather than as simply asserting a rule of distinction between parties to the dispute. Herlihy quotes Thomas Kuhn's observation that "one of the strongest, if still unwritten, rules of scientific life is the prohibition of appeals to heads of state *or to the populace at large* in matters scientific."[23] But is this changing? The Arkansas case forces us to consider the implications of the *legal* recognition of the unique status of science as exclusive arbiter of its own professional status and achievement. If science continues to look to the courts for "protection" of its curriculum or for a consensus definition of science, will its position within society as a neutral and independent arbiter in social disputes be weakened?

The scientific community has been accustomed to unquestioned control over the rules, members, and boundaries of the research system;

but, in determining the content of science education, scientists confront the reality of public school systems that consolidate effort and resources for the common good. Rather than each employing tutors suited to our specific political, religious, and moral beliefs, we join with other parents in the community and pay for a common system. Society also decides that, for the greater benefit of all, certain subjects must be taught and other subjects not. And underlying all of this is the belief by all parties that, whatever else happens in the classroom, teaching must be intellectually honest and fair.

Conflict arises because the creationists seek equal status in the public schools for ideas embracing belief in a (sectarian) Creator, a young earth, and a single creation of all living kinds at one time. Their definition of "fair play" demands "unbiased presentation" and "balanced treatment" of creation-science. To the scientist, however, *any* discussion of creation-science seems unfair because, measured by the accepted indicators of scientific quality, it is intellectually dishonest. A *Science* magazine editorial on the *McLean* decision declared, for example, that the creationists lost because "truth is not on their side."[24] Both "sides" feel beleaguered and justified in their attacks. (This reaction often leads to a type of hostile and loaded language that we have striven to avoid in this book.) But, as Dudley Field Malone argued during the Scopes trial, "there is no more justification for imposing the conflicting view of the Bible on courses of biology than there would be for imposing the views of biologists on courses of comparative religion."[25]

The Arkansas case suggests the need for a fresh look at the appropriate roles for both religion and science in modern society. Gilkey accuses those scientists who hold to the "myth" that religion will simply vanish in a secular and scientific culture of being short-sighted and ill-equipped to meet the creationists' challenge to the very foundations of public support of science. For, to build their case for placing "creation-science" in the public schools, the creationists need the authority of science. By bringing scientists as witnesses to testify in their court cases and by using scientific evidence to shore up their theories, the creationists use science's own reputation to gain public acceptance for theories with which, in fact, the majority of scientists disagree. In the Creationism debate, science's carefully maintained privilege and legitimizing status within American society has been turned back on science in a public forum.

References

1. "Law on Creation Theory Proves Boon for Governor of Arkansas,"*New York Times* (22 March 1981): A26. The *Times* reported that "by signing and praising the bill within hours of its reaching his desk. . . , and without even reading it, . . . Mr. White reinforced his standing with conservative and evangelical groups."

2. *McLean v. Arkansas*, 529 F.Supp. 1255 (Eastern District, Arkansas 1982).

3. This discussion of the development of Creationism relies on Judge Overton's opinion in *McLean*, the Plaintiffs' Pre-Trial Brief in *McLean*, the essay by Dorothy Nelkin in this book, and Dorothy Nelkin, *Science Textbook Controversies and the Politics of Equal Time* (Cambridge, MA: MIT Press, 1977).

4. *State of Tennessee v. John Thomas Scopes*, Nos. 5231 and 5232, Circ. Ct. Rhea County, Tennessee (1925); *Scopes v. State*, 154 Tennessee 105, 289 S.W. 363 (1927).

5. Quoted in "Creation vs. Evolution—Battle in the Classroom" a television documentary film produced by KPBS-TV (San Diego, 1982), "Synthesis III" series (producer: Ryall Wilson).

6. Letter to the Editor, *Arkansas Democrat* (18 December 1981).

7. Philip Kitcher, *Abusing Science: The Case Against Creationism* (Cambridge, MA: MIT Press, 1982), p. 186.

8. Russell Doolittle, quoted in "Putting Darwin Back Into the Dock," 117 *Time* (16 March 1981): 82.

9. Quoted in Gary Crawford's chapter, this volume.

10. Mark Herlihy makes these points in his chapter in this volume.

11. *Epperson v. Arkansas*, 393 U.S. 97 (1968). All quotations in this paragraph are from the Supreme Court opinion in *Epperson*.

12. *Ibid.*, pp. 103–104.

13. *Daniel v. Waters*, 515 F.2d 485 (1975). All quotations in this paragraph are from the District Court opinion in *Daniel*.

14. Abbreviated paraphrase of Justice Black's opinion in *Everson v. Board of Education*, 330 U.S. 1, 15–16 (1947).

15. *Lemon v. Kurtzman*, 403 U.S. 602 (1971).

16. A point made by Eric Holtzman and David Klasfeld, this volume.

17. Judge Overton quotes one such oath in footnote 7 to his Opinion in *McLean v. Arkansas*.

18. A point made in Daniel Yankelovich, "Changing Public Attitudes to Science and the Quality of Life," 7 *Science, Technology, & Human Values* (Spring 1982): 23–29; and in Kenneth Prewitt, "The Public and Science Policy," 7

Marcel C. La Follette   14

Science, Technology, & Human Values 39 (Spring 1982):5–14; see also *Quality in Science*, M. C. La Follette, ed. (Cambridge, MA: MIT Press, 1982).

19. "Opinion Roundup—School Prayer: Overwhelming Endorsement," *Public Opinion* (June/July 1982): 40.

20. "Opinion Roundup," *Public Opinion* (October/November 1981):23–26.

21. *Ibid.*, p. 24.

22. S. J. Holmes, quoted in Nelkin, this volume.

23. Thomas S. Kuhn, *The Structure of Scientific Revolutions*, second edition (Chicago: University of Chicago Press, 1971), p. 168, italics mine.

24. Editorial, 215 *Science* (8 January 1982). In summarizing the *McLean* trail, a reporter for the *Arkansas Gazette* commented: "Some ACLU witnesses [scientists] attacked creationist writings as not only bad science—poorly researched or thought out—but also as being outright 'dishonest.'" (George Wells, "Creationism Case Lacked Excitement of 'Monkey Trial,'" *Arkansas Gazette* (20 December 1981), p. 9A.

25. Dudley Field Malone, in *State of Tennessee v. John Thomas Scopes*, Nos. 5231 and 5232, Circ. Ct. Rhea County, Tennessee (1925), p. 293 of the transcript. [*Monkey Trial*, Sheldon Norman Grebstein, ed. (Boston: Houghton Mifflin, 1960), p. 93.]

Act 590 of 1981

General Acts, 73rd General
Assembly, State of
Arkansas

AN ACT TO REQUIRE BALANCED TREATMENT OF CREA-
TION-SCIENCE AND EVOLUTION-SCIENCE IN PUBLIC
SCHOOLS; TO PROTECT ACADEMIC FREEDOM BY PROVID-
ING STUDENT CHOICE; TO ENSURE FREEDOM OF RELIGIOUS
EXERCISE; TO GUARANTEE FREEDOM OF BELIEF AND
SPEECH; TO PREVENT ESTABLISHMENT OF RELIGION; TO
PROHIBIT RELIGIOUS INSTRUCTION CONCERNING
ORIGINS; TO BAR DISCRIMINATION ON THE BASIS OF CRE-
ATIONIST OR EVOLUTIONIST BELIEF; TO PROVIDE DEFI-
NITIONS AND CLARIFICATIONS; TO DECLARE THE
LEGISLATIVE PURPOSE AND LEGISLATIVE FINDINGS OF
FACT; TO PROVIDE FOR SEVERABILITY OF PROVISIONS; TO
PROVIDE FOR REPEAL OF CONTRARY LAWS; AND TO SET
FORTH AN EFFECTIVE DATE.

Be It Enacted by the General Assembly of the State of Arkansas:

SECTION 1. *Requirement for Balanced Treatment.* Public Schools
within this State shall give balanced treatment to creation-science and
to evolution-science. Balanced treatment to these two models shall be
given in classroom lectures taken as a whole for each course, in textbook
materials taken as a whole for each course, in library materials taken
as a whole for the sciences and taken as a whole for the humanities,
and in other educational programs in public schools, to the extent that
such lectures, textbooks, library materials, or educational programs
deal in any way with the subject of the origin of man, life, the earth,
or the universe.

SECTION 2. *Prohibition against Religious Instruction.* Treatment of either evolution-science or creation-science shall be limited to scientific evidence for each model and inferences from those scientific evidences, and must not include any religious instruction or references to religious writings.

SECTION 3. *Requirement for Nondiscrimination.* Public schools within this State, or their personnel, shall not discriminate, by reducing a grade of a student or by singling out and making public criticism, against any student who demonstrates a satisfactory understanding of both evolution-science and creation-science and who accepts or rejects either model in whole or part.

SECTION 4. *Definitions.* As used in this Act:

(a) "Creation-science" means the scientific evidences for creation and inferences from those scientific evidences. Creation-science includes the scientific evidences and related inferences that indicate: (1) Sudden creation of the universe, energy, and life from nothing; (2) The insufficiency of mutation and natural selection in bringing about development of all living kinds from a single organism; (3) Changes only within fixed limits of originally created kinds of plants and animals; (4) Separate ancestry for man and apes; (5) Explanation of the earth's geology by catastrophism, including the occurrence of a worldwide flood; and (6) A relatively recent inception of the earth and living kinds.

(b) "Evolution-science" means the scientific evidences for evolution and inferences from those scientific evidences. Evolution-science includes the scientific evidences and related inferences that indicate: (1) Emergence by naturalistic processes of the universe from disordered matter and emergence of life from nonlife; (2) The sufficiency of mutation and natural selection in bringing about development of present living kinds from simple earlier kinds; (3) Emergence by mutation and natural selection of present living kinds from simple earlier kinds; (4) Emergence of man from a common ancestor with apes; (5) Explanation of the earth's geology and the evolutionary sequence by uniformitarianism; and (6) An inception several billion years ago of the earth and somewhat later of life.

(c) "Public schools" mean secondary and elementary schools.

SECTION 5. *Clarification.* This Act does not require or permit instruction in any religious doctrine or materials. This Act does not require any instruction in the subject of origins, but simply requires instruction in both scientific models (of evolution-science and creation-science) if public schools choose to teach either. This Act does not require each individual textbook or library book to give balanced treatment to the models of evolution-science and creation-science; it does not require any school books to be discarded. This Act does not require each individual classroom lecture in a course to give such balanced treatment, but simply requires the lectures as a whole to give balanced treatment; it permits some lectures to present evolution-science and other lectures to present creation-science.

SECTION 6. *Legislative Declaration of Purpose.* This Legislature enacts this Act for public schools with the purpose of protecting academic freedom for students' differing values and beliefs; ensuring neutrality toward students' diverse religious convictions; ensuring freedom of religious exercise for students and their parents; guaranteeing freedom of belief and speech for students; preventing establishment of Theologically Liberal, Humanist, Nontheist, or Atheist religions; preventing discrimination against students on the basis of their personal beliefs concerning creation and evolution; and assisting students in their search for truth. This Legislature does not have the purpose of causing instruction in religious concepts or making an establishment of religion.

SECTION 7. *Legislative Findings of Fact.* This Legislature finds that:

(a) The subject of the origin of the universe, earth, life, and man is treated within many public school courses, such as biology, life science, anthropology, sociology, and often also in physics, chemistry, world history, philosophy, and social studies.

(b) Only evolution-science is presented to students in virtually all of those courses that discuss the subject of origins. Public schools generally censor creation-science and evidence contrary to evolution.

(c) Evolution-science is not an unquestionable fact of science, because evolution cannot be experimentally observed, fully verified, or logically falsified, and because evolution-science is not accepted by some scientists.

(d) Evolution-science is contrary to the religious convictions or moral values or philosophical beliefs of many students and parents, including individuals of many different religious faiths and with diverse moral values and philosophical beliefs.

(e) Public school presentation of only evolution-science without any alternative model of origins abridges the United States Constitution's protections of freedom of religious exercise and of freedom of belief and speech for students and parents, because it undermines their religious convictions and moral or philosophical values, compels their unconscionable professions of belief and hinders religious training and moral training by parents.

(f) Public school presentation of only evolution-science furthermore abridges the Constitution's prohibition against establishment of religion, because it produces hostility toward many Theistic religions and brings preference to Theological Liberalism, Humanism, Nontheistic religions, and Atheism, in that these religious faiths generally include a religious belief in evolution.

(g) Public school instruction in only evolution-science also violates the principle of academic freedom, because it denies students a choice between scientific models and instead indoctrinates them in evolution-science alone.

(h) Presentation of only one model rather than alternative scientific models of origins is not required by any compelling interest of the State, and exemption of such students from a course or class presenting only evolution-science does not provide an adequate remedy because of teacher influence and student pressure to remain in that course or class.

(i) Attendance of those students who are at public schools is compelled by law, and school taxes from their parents and other citizens are mandated by law.

(j) Creation-science is an alternative scientific model of origins and can be presented from a strictly scientific standpoint without any religious doctrine just as evolution-science can, because there are scientists who conclude that scientific data best support creation-science and because scientific evidences and inferences have been presented for creation-science.

(k) Public school presentation of both evolution-science and creation-science would not violate the Constitution's prohibition against establishment of religion, because it would involve presentation of the scientific evidences and related inferences for each model rather than any religious instruction.

(l) Most citizens, whatever their religious beliefs about origins, favor balanced treatment in public schools of alternative scientific models of origins for better guiding students in their search for knowledge, and they favor a neutral approach toward subjects affecting the religious and moral and philosophical convictions of students.

SECTION 8. *Short Title.* This Act shall be known as the "Balanced Treatment for Creation-Science and Evolution-Science Act."

SECTION 9. *Severability of Provisions.* If any provision of this Act is held invalid, that invalidity shall not affect other provisions that can be applied in the absence of the invalidated provisions, and the provisions of this Act are declared to be severable.

SECTION 10. *Effective Date.* The requirements of the Act shall be met by and may be met before the beginning of the next school year if that is more than six months from the date of enactment, or otherwise one year after the beginning of the next school year, and in all subsequent school years.

Signed on 19 March 1981 by Governor Frank White.

The Legal Arguments:
Excerpts from the Plaintiffs'
Preliminary Outline and
Pre-Trial Brief

[*Editor's note*. During the course of litigation, all parties to a suit submit documents to the court outlining their arguments and the legal proofs on which they rest. In *McLean*, the formal briefs were well-written, closely argued, and very long. To present some of the important arguments for both sides, in their own words, we present here and in the following chapter heavily edited versions of the "Preliminary Outline of the Legal Issues and Proof" filed with the Arkansas court by both plaintiffs and defendants. Excerpts from the trial briefs are also included. Each document contains a hundred or more legal citations, often multiple, to precedents or pre-trial document exhibits, which have been omitted here. A list of the cases cited is included at the end of the following chapter (see p. 44). The following guide should help readers unfamiliar with legal citations:

1. *Everson v. Board of Education*, 330 U.S. 1, 15–16 (1947), refers to a 1947 U.S. Supreme Court decision found in Volume 330 of *U.S. Reports*. The case begins on page 1, and specific reference is to pages 15 through 16.
2. *Everson* at 16 refers to page 16 in the *Everson* decision.
3. *Malnak v. Yogi*, 440 F. Supp. 1284, 1322 (D. N.J. 1977); *affirmed per curiam*, 592 F.2d 197 (3rd Circuit, 1979). This citation shows the history of the litigation. The first opinion was handed down in 1977 by the Federal District Court for New Jersey and published in Volume 440 of the *Federal Supplement*. The case then was appealed to the U.S. Court of Appeals for the Third Circuit, and the decision affirming the lower court's decision was published in Volume 592 of the second series of the *Federal Reporter*.]

PLAINTIFFS' OUTLINE OF LEGAL ISSUES AND PROOF*

Overview

This action challenges the constitutionality of the Arkansas "Balanced Treatment for Creation-Science and Evolution-Science Act." . . . The evidence will show that Act 590 (1) violates the Establishment Clause of the First Amendment to the Constitution of the United States; (2) unconstitutionally abridges academic freedom by injecting the orthodoxy of a state-sponsored theory of science into public school classrooms; and (3) violates the rights of Arkansas teachers under the Due Process Clause of the Fourteenth Amendment by its vagueness.

Factual Background

Act 590 was sponsored by Senator James L. Holsted, a religious fundamentalist, who has publicly stated that the bill embodies his religious beliefs. Act 590, including lengthy "legislative findings of fact," is textually identical to a "Model Bill" drafted by Paul Ellwanger, founder and chairman of the "Citizens for Fairness in Education," who is engaged in a campaign opposing "the use of tax dollars . . . against God." . . . [citations to pre-trial exhibitions and to Ellwanger deposition omitted]. Ellwanger sought and received advice on his Model Bill from Wendell Bird, now staff counsel for the Institute for Creation Research (an affiliate of Christian Heritage College), and from other individuals believed to be associated with "creationist organizations" . . . [citation to Ellwanger deposition omitted].

The Arkansas Department of Education was not consulted before passage of Act 590. . . . [citation to depositions from public officials omitted] . . . Moreover, Act 590 is the only known instance where the legislature has given specific definition to an academic subject matter and prescribed when, how, and in what circumstances it shall be taught. . . . [legal citations omitted]

Act 590 posits two opposing "scientific models": one of "creation-science," the other of "evolution-science" (Act 590, Section 4). The evidence will show that the six statutory elements of the "creation-science" model are derived directly from the Biblical account of creation, literally interpreted. Act 590 requires that whenever "the subject of

*McLean v. Arkansas, U.S. District Court, Eastern District of Arkansas, Western Division, Civil Action LRC 81-322, Plaintiffs' Preliminary Outline of Legal Issues and Proof.

the origin of man, life, the earth, or the universe" is dealt with in any way in a course, "balanced treatment"—a term nowhere defined in Act 590—must be given to both models (Act 590, Section 1).

Outline of Argument

I. Act 590 Establishes Religion
The Establishment Clause forbids government to "pass laws which aid one religion, aid all religions, or prefer one religion over another" (*Everson v. Board of Education* at 15). Act 590 violates the Establishment Clause under each aspect of the three-prong test developed by the Supreme Court: "First, the statute must have a secular legislative purpose; second, its principal or primary effect must be one that neither advances nor inhibits religion. . . ; finally, the statute must not foster 'an excessive government entanglement with religion' " (*Lemon v. Kurtzman* at 612–613).

A. Act 590 Reflects a Religious, Rather Than a Secular, Purpose   Act 590 is unconstitutional for the same reasons the Supreme Court declared invalid, in *Epperson v. Arkansas*, the Arkansas statute that prohibited public school instruction about evolution. Even though the statute did not expressly mention any religious beliefs, the Supreme Court held the statute unconstitutional under the Establishment Clause because its sole purpose was to "blot out a particular theory because of its supposed conflict with the Biblical* account, literally read."

The Court stressed that public school education cannot be tailored to conform to the beliefs of any religion: "There is and can be no doubt that the First Amendment does not permit the State to require that teaching and learning must be tailored to the principles or prohibitions of any religious sect or dogma. . . . This prohibition is absolute. It forbids alike the preference of a religious doctrine or the prohibition of a theory which is deemed antagonistic to a particular dogma."

The Declaration of Purpose in Section 6 of Act 590 belies any secular underpinnings for the Act . . . [footnote omitted] . . . The stated purposes—chief among which are protecting students' religious beliefs, ensuring "free exercise" of religion, and preventing "establishment" of

---

*The excerpts from the defendants' and plaintiffs' briefs retain their original capitalizations—for example, the use of "State" and "biblical" by the defendants and "state" and "Biblical" by the plaintiffs.—Ed.

"Theologically Liberal, Humanist, Nontheist, or Atheist religions" . . . [footnote omitted] . . . —show conclusively that religious
considerations prompted the statute. See also Section 7(d) and (e) of
Act 590, where the legislature worries that evolution is "contrary to
the religious convictions" of students and parents and "undermines
their religious convictions."

Moreover, the source of Act 590 and the events immediately preceding
its passage evidence a religious purpose . . . [footnote omitted] . . . For
example, draftsman Ellwanger encourages "Christian political action"
in support of his Model Bill . . . [citations to exhibit and deposition
omitted] . . . In a recent letter seeking contributions "to carry on in
getting the other 48 states to successfully follow the winning strategy
we suggested for Arkansas and Louisiana," Ellwanger emphasized that,
"this legislative process [is being used] to help expose and neutralize
such gross errors as Humanism, for without evolution it and all God-
less belief systems would simply crumble." . . . [citation to exhibit
omitted] . . .

The evidence also will show that fundamentalist religious groups and
individuals secured the introduction and passage of Act 590. The Act's
sponsor, Senator Holsted, has admitted that he cannot separate Act
590 from his belief in a Creator. . . .

Finally, the proof will show that Arkansas has a long history of
legislation designed to protect the beliefs of Biblical literalists and to
promote religious faith and observance [legal citation omitted].

In sum, Act 590 is the legislative enactment of a religious antipathy
to evolution. The Act does not reflect a clear secular purpose but instead
exists only because of "fundamentalist sectarian conviction" (*Epperson
v. Arkansas* at 108). . . .

B. The Primary Effect of Act 590 Is to Advance Religion   The Establishment Clause is violated by a statute which has the direct and immediate effect of advancing religion. . . . Courts have been particularly
sensitive to the use of public schools for the promotion of religious
beliefs. Not only have overt religious exercises, such as prayer, Bible
readings, and religion lessons, been held unconstitutional . . . [legal citations omitted] . . . but more subtle attempts to advance essentially
religious beliefs through public school instruction, or to modify public
school curricula to conform with religious tenets, have been invalidated
as well. (*Stone v. Graham*, posting of Ten Commandments in classrooms; *Epperson v. Arkansas*, prohibiting the teaching of evolution;

Malnak v. Yogi, teaching the science of creative intelligence/transcendental meditation.)

1. Instruction in "Creation-Science" Advances Religion   The evidence will prove that "creation-science" inculcates religious beliefs, particularly beliefs which derive solely from a literal reading of the Bible. Promoted by a recognizable group of religious fundamentalists, "creation-science" is designed to propagate their religious views. The Institute for Creation Research, the preeminent "creation-science" organization in the country, says it is:

engaged in an active threefold ministry—research, writing, and teaching—in the field of scientific Biblical creationism. We are convinced this is the most effective way in which recognition of God as a sovereign Creator and Savior can be restored in our modern world, especially for the multitudes of young people in our schools who have been indoctrinated for so long with the false and harmful philosophy of evolutionary humanism [Henry Morris, letter to new subscribers of *Acts & Facts* (undated).]

Similarly, two leaders of the Creation-Science Research Center ("CSRC") describe their purpose in writing one of the CSRC's major "creation-science" texts as follows:

[T]he Bible also says that the Bible and creation are to be received as true by faith (Hebrews 11:30). Nevertheless we are commanded to persuade men (2 Corinthians 5:11), using every reasonable means. . . . [Robert E. Kofahl and Kelly L. Segraves, *The Creation Explanation: A Scientific Alternative to Evolution*, Creation-Science Research Center (1975), hereinafter cited as *The Creation Explanation*.]

. . . Teaching "creation-science" advances religion because it invokes and relies upon the existence and activity of a creator (a concept that necessarily imports a Supreme Being) . . . [legal citations omitted] . . . and reflects not only a religious view of origins, but one which is peculiar to religious fundamentalists. Similar state attempts to teach religious beliefs under the guise of secular labels such as science, philosophy, or literature have been held unconstitutional. . . .

"Creation-Science" works often speak not of "God," but of a "Creator," and describe not a "divine creation" but a "special creation." . . . [legal citations omitted] . . . Nonetheless, a "Creator" is indistinguishable from a "Supreme Being" or "God." (See *Malnak v. Yogi.*) The effect of teaching such doctrines is to destroy the neutrality the Establishment Clause commands and to give preference to certain religious views.

2. Instruction in "Creation-Science" Advances No Secular Educational Goal Because "Creation-Science" Is Not Science   Plaintiffs' proof will show that "creation-science," rather than being an alternative scientific theory to evolution, as asserted by Act 590, is in reality religious apologetics. See, for example, *The Creation Explanation*, p. 181 (referring to "the new science-based apologetic for biblical creation"). Even stripped of all explicit Biblical references, "creation-science" remains a body of factual inferences specifically designed to buttress belief in a literal interpretation of Genesis—not a secular body of conclusions reached by applying the ordinary standards of scientific discourse. "Creation-science" is admitted to be an exercise in evaluating facts according to a particular religion, ignoring or rejecting those facts which are viewed as contradictory:

While, as scientists, creationists must study as objectively as possible the actual data of geology, as Bible-believing Christians, we must also *insist that these be correlated within the framework of Biblical revelation. This restriction requires* rejection of the traditional uniformitarian approach [of geology] . . . [Henry M. Morris, "Diversity of Opinions Found in Creationism," 11 *Creation Research Society Quarterly* 3 (December 1974): 173 (emphasis added)].

The Christian student of origins approaches the evidence from geology and paleontology with the biblical record in mind, *interpreting that evidence in accord with the facts divinely revealed in the Bible. . . .[The Creation Explanation*, p. 40 (emphasis added)].

Bible-believing students of the biological sciences possess *a guide for their interpretation of the available data*, the biblical record of divine creation contained in Genesis . . . [*The Creation Explanation*, p. 69 (emphasis added)].

. . . [C]reationists . . . start with divine revelation and only then address the facts. . . . Science concerns itself solely with natural phenomena and natural processes, eschewing resort to the supernatural, whereas "creation-science" relies upon, espouses, and is dependent upon supernatural processes. A key feature of science is that it is *always* somewhat tentative and *always* subject to revision in the light of new knowledge. "Creation-science," in contrast, is based on a claim of absolute truth, and is not subject to revision in the light of new knowledge.

As the proof will show, "creation-science" does not follow the scientific method, and the "scientific evidences" on which "creation-science" relies are not scientific at all. Among other things, "creation-science" ignores, distorts and fails to take account of relevant data; it relies upon out-of-date and thoroughly discredited data and authorities; and it ignores, distorts and rewrites scientific principles, solely to give the appearance of support for pre-determined conclusions.

C. Act 590 Creates Excessive Entanglement Between the State and Religion   "Excessive entanglement" between religion and government will invalidate a statute under the Establishment Clause (*Lemon v. Kurtzman* at 612–613). This may be *administrative* entanglement (continuing state vigilance to prevent the promotion of religion); or *political* entanglement (the potential for dividing the electorate on the basis of voters' religious beliefs). Both types of entanglement exist here.

Act 590, Section 2, provides that "Treatment of either evolution-science or creation-science . . . *must not include any religious instruction or references to religious writings*" (emphasis added). Yet, the religious concepts inherent in "creation-science" are so pervasive that drawing distinctions between "religious" and "nonreligious" discourse in the classroom and in textbooks will necessarily entwine state officials in the kinds of delicate religious judgments barred by the administrative entanglement test. . . .

The Supreme Court has pointed out that, however entangling judgments on textbooks may be, the necessary monitoring of classroom discussions is yet more troubling. *Lemon v. Kurtzman* invalidated a law permitting state purchase of secular educational services from nonpublic schools, including parochial schools. The Supreme Court held the statute unconstitutional because the monitoring necessary to enforce the statute's restrictions against religious instruction would hopelessly entwine the State and religion. . . .

Lower Federal courts have applied the concept of administrative entanglement to invalidate not only state involvement in private religious schools, but also religious intrusions in public institutions. In *Brandon v. Board of Education*, the court held that use of public school facilities for prayer meetings would involve excessive entanglement:

[S]urveillance would be required to insure that attendance and participation in the meetings remained voluntary and that no pressure, no matter how subtle, was placed on other students to coerce involvement [*Brandon* at 1230].

. . . [additional legal citations omitted] . . .

The "political entanglement" test examines the "divisive political potential" of legislation relating to religious questions. The Supreme Court has struck down state legislation assisting church-related schools in part on the ground that such assistance engenders unhealthy political disputes between religious groups. In *Lemon v. Kurtzman,* the Court noted:

Ordinarily political debate and division, however vigorous or even partisan, are normal and healthy manifestations of our democratic system of government, but political division along religious lines was one of the principal evils against which the First Amendment was intended to protect. . . . The potential divisiveness of such conflict is a threat to the normal political process. [*Lemon* at 622.]

See also *Stone v. Graham* (a Kentucky statute requiring the posting of a copy of the Ten Commandments in public classrooms held unconstitutional partly because of its potential for political divisiveness), and *Hall v. Bradshaw* ("motorist's prayer" on North Carolina map published and distributed by State Department of Transportation held unconstitutional partly because of its potential for political divisiveness).

Act 590 has set loose in Arkansas (as in the nation) a storm of religious acrimony. Neighbor has been set against neighbor as Arkansas citizens divide on the issue of teaching "creation-science," largely on the basis of their religious beliefs. Political entanglement of this nature in religious controversies threatens the very foundations of American democratic pluralism and, because of the serious threat it poses, is unconstitutional under the Establishment Clause.

II. Act 590 Unconstitutionally Abridges Academic Freedom Rights

Academic freedom is a constitutionally protected right which emanates from the First and Fourteenth Amendments. "Our Nation is deeply committed to safeguarding academic freedom, which is of transcendent value to all of us. . . . That freedom is therefore a special concern of the First Amendment, which does not tolerate laws that cast a pall of orthodoxy over the classroom" [*Kevishian v. Board of Regents* at 603]. . . . [additional legal citations omitted] . . .

Academic freedom, although not defined in its broadest contours, at least embraces the right of the teacher to teach and the right of the student to learn. The Eighth Circuit has been said to be in accord with opinions which find, for example, that public school teachers have at least "the substantive right . . . to choose a teaching method which in the court's view serve[s] a demonstrated educational purpose" (*Webb v. Lake Mills Community School District* at 799) . . . [additional legal citations omitted] . . .

Since the ultimate goal of academic freedom is education, it limits the power of the state to disrupt the ordinary and proper functioning of the educational system. A state, notwithstanding its otherwise broad control over the operation of its public schools, cannot disrupt the educational process with "unusual or irregular intervention" (*Pico v. Board of Education* at 414). . . . A state may not arbitrarily prescribe curriculum content that inculcates propaganda or orthodoxy on public school children or has the effect of censoring or chilling academic discussion (See *Epperson v. Arkansas* at 105). . . .

Act 590 impermissibly interferes with academic freedom by censoring evolution to the extent it is not "balanced" with "creation-science." . . . The evidence will show that, because of the Act, many teachers will choose to teach neither evolution nor "creation-science." This chilling effect on classroom instruction will effectively deprive many schoolchildren—who are in school under the coercion of compulsory education laws—of their right to learn about evolution, thus achieving indirectly what cannot be achieved directly under the Establishment Clause (See *Epperson v. Arkansas*).

Moreover, to the extent that evolution and "creation-science" *are* taught, the strictures of Act 590 may make teachers reluctant to evaluate either model, to state professional opinions or to answer students' questions. . . .

Furthermore, in passing Act 590, the Arkansas legislature has overridden the professional judgment of teachers, scientists and educators, that "creation-science" lacks recognized educational value, thereby giving this subject matter a most peculiar and privileged place in the curriculum. Act 590 thus casts a "pall of orthodoxy" over the public school classroom by mandating instruction in a state-approved doctrine. . . .

Finally, Act 590's "irregular intervention" in the process of education makes mere schoolchildren the arbiters of the validity of the competing concepts of "creation-science" and evolution. . . . By mandating that they be taught about two alternative models, Act 590 also foists upon them a false—and painful—dichotomy: because "creation-science" insists that a belief in creation is inconsistent with evolution, children may be led to think they must choose between religion and science. The Constitutional doctrine of academic freedom stands as a bulwark against this kind of distortion of the educational process.

III. Act 590 Is Unconstitutionally Vague

It has long been the law that "[A] statute which either forbids or requires the doing of an act in terms so vague that men of common intelligence must necessarily guess at its meaning and differ as to its application, violates the first essential of due process of law" (*Connally v. General Construction Company* at 391). Such a statute is void for vagueness under the Due Process Clause.

This test applies to civil statutes, as well as to criminal laws, where liberty or property interests are affected by the legislation. Thus courts routinely analyze civil statutes to determine whether they are unconstitutionally vague. . . .

Under the Arkansas Teacher Fair Dismissal Act of 1979, every Arkansas public school teacher has a property right in continued employment during the term of his or her annual contract. . . . Moreover, every non-probationary teacher (i.e., every teacher who has been continuously employed by a particular school district for three years) has a property right in successive renewals of his or her annual contract. . . . No Arkansas teacher's contract may be terminated and no non-probationary teacher's contract may be non-renewed except for cause which is "not arbitrary, capricious, or discriminatory." . . . In addition to a property right in continued employment, a public school teacher has a liberty right in the maintenance of his or her professional reputation. . . .[legal citations omitted] . . . These liberty and property rights may be seriously jeopardized by Act 590, yet teachers "must necessarily guess at its meaning," and local school boards must necessarily "differ as to its application" (*Connally v. General Construction Co.* at 391).

Act 590 is hopelessly ambiguous in numerous respects. Among other things, although it mandates "balanced treatment," it does not define the term, nor does it state whether, in the face of a legislative finding that "creation-science" is an "alternative *scientific* model," (Act 590, Section 7.j, emphasis added), a teacher remains free to state his professional view that "creation-science" is not scientific at all.

Even the drafter and the sponsor . . . [footnotes omitted] . . . of Act 590 do not seem to know precisely what conduct is regulated by Act 590, nor do they seem to know with certainty whether speech is regulated and, if so, how speech is regulated. How much more confused will be a teacher charged with compliance with Act 590—possibly under penalty of loss of job or his or her professional reputation—or a local school board charged with enforcement of Act 590?

Act 590 does not "give the person of ordinary intelligence a reasonable opportunity to know what is prohibited, so that he may act accordingly," with the unconstitutional consequence that it "may trap the innocent by not providing fair warning" (*Gravned v. City of Rockford* at 108). . . . "Balanced treatment" may be construed under the Act as "equal time," or it may require some proportional allotment of attention, or it may, as Ellwanger believes, require "unbiased" treatment. Similarly, the scope of the Act is uncertain; it could be deemed to prohibit teachers from expressing either opinions or informed judgments regarding either evolution or "creation-science" or both. The local school boards of Arkansas may apply Act 590 differently, and discriminatorily, perhaps depending on the local political climate.

. . . The potential Act 590 holds for exerting a "chilling effect"—for encouraging teachers, for example, to refrain from teaching evolution at all so as to avoid having to teach "creation-science" as science, or to refrain from expressing their views on "creation-science" as science— is abhorrent to the Due Process Clause. . . . Act 590 is a vague statement, which must be held unconstitutional.

Conclusion
For the legal reasons set forth herein and on the evidence to be presented at trial, plaintiffs will ask the Court to declare Act 590 unconstitutional under the First and Fourteenth Amendments to the United States Constitution and to enter a permanent injunction against its enforcement.

EXCERPTS FROM PLANTIFFS' PRE-TRIAL BRIEF*

[*Editor's note*. In the pre-trial brief, plaintiffs' attorneys expanded upon their argument that the introduction of creation-science into the public school classroom "distorts the educational system for religious ends."]

. . . Clearly, Act 590 is a unique legislative intrusion into the classroom. Its effects are devastating as far as the subject of biology is concerned, for evolution is "the organizing principle of biology" (Ernst Mayr, "Evolution," *Evolution* 3, *Scientific American*, 1978; see also J. Scow, "The Genesis of Equal Time," 2 *Science 81* 54, December 1981). Many

---

*Excerpted from *McLean v. Arkansas*, U.S. District Court, Eastern District of Arkansas, Western Division, Civil Action LRC 81-322, Plaintiffs' Pre-Trial Brief.

biology teachers feel they cannot effectively teach their subject without reference to evolution, yet the only alternative under Act 590 is that they give "balanced treatment" . . . to the religious, nonscientific doctrine of "creation-science."

Nor is biology the only affected discipline; all of science is drawn into the religious net of "creation-science":

Adoption of creationist "theory" requires, at a minimum, the abandonment of essentially all of astronomy, much of modern physics, and most of the earth sciences. Much more than evolutionary biology is at stake. (Allen Hammond and Lynn Margulis, "Farewell to Newton, Einstein, Darwin. . . ," 2 *Science 81* 55, December 1981.)

How, for example, can a chemistry teacher teach the observed decay of elements in the periodic table, having half-lives of billions of years, and still give "balanced treatment" to "creation-science," which posits that nothing can be older than about 10,000 years?

History, sociology and numerous other subjects are likewise affected. Imagine the dilemma of a history teacher, for example, who wants to teach about the *Scopes* trial as an historical event, yet must give "balanced treatment" to "creation-science."

Even defendants' witnesses acknowledge that the result of Act 590 will be something other than effective teaching and educational clarity:

Q. If the only evidence in support of Creation-Science are the same evidences which also support panspermiogenesis, or any other Theory of Origin, other than evolution, how does the school child understand that those are evidences in support of Creation Science?

A. *The school child is going to experience a considerable amount of confusion on these topics.* (Deposition by W. Scot Morrow at p. 144, emphasis aded.) . . .

[*Editor's note.* Later in the pre-trial brief, the plaintiffs asserted that "Act 590 not only freezes scientific definitions, but also requires to be taught as science a view that has not proven itself and gained acceptance within the scientific community, thus creating special dangers for the academic freedom." The brief then derives some specific examples:]

. . . When a teacher chooses, under Act 590, to present evolution and "creation-science," the Act's strictures may render teachers reluctant to evaluate either model, to state professional opinions, or to answer questions, thereby chilling the exercise of First Amendment academic freedom rights. In *Moore v. Gaston County Board of Education*, a teacher's inadvertent use of the word "evolution" in a history lesson triggered student questions and a discussion of the teacher's views on Darwinism. Concluding that the teacher's dismissal for this incident abridged both academic freedom rights and the Establishment Clause the court noted:

> If a teacher has to answer searching, honest questions only in terms of the lowest common denominator of the professed beliefs of those parents who complain loudest, this means that the state through the public schools is impressing the particular religious orthodoxy of those parents upon the religious and scientific education of the children by force of law. (*Idem* at 1043.)

Conversely, to maintain their professional integrity and to avoid the risk of discipline or dismissal under Act 590, many teachers will choose to teach neither evolution nor "creation-science," thus depriving students of the opportunity to learn major portions of the curriculum. A state could not, without violating the Establishment Clause, insulate religious beliefs by proscribing the teaching of evolution (*Epperson v. Arkansas*). Yet through Act 590, Arkansas will succeed in imposing upon students as science its orthodox view of origins or, alternatively, in censoring the opposing view.

The Constitution's prohibition of statutorily mandated instruction in "creation-science" is not in any way censorship. "Creation-science" proponents are not barred from the development and communication of creationist theories outside the classroom, nor from attempting to achieve equal status in the marketplace of scientific ideas with those established scientific concepts now included in the public school curriculum. The Constitution simply safeguards academic freedom by forbidding the state to shortcircuit the educational decision-making process or to confer academic status upon a body of knowledge for reasons unrelated to the educational merits of that body of knowledge. . . .

The Legal Arguments:
Excerpts from the Defendants' Preliminary Outline and Pre-Trial Brief

[*Editor's note*. See the editor's note at the beginning of the preceding chapter (p. 20). Citations for the cases cited in both chapters are given on p. 44.]

DEFENDANTS' OUTLINE OF THE LEGAL ISSUES AND PROOF*

Overview

This case presents issues of first impression. The Balanced Treatment for Creation-Science and Evolution-Science Act . . . is designed to ensure that if a school teaches theories of origin, that it will do so in a neutral manner which will further academic freedom. The central issue in this case is whether a State-mandated, neutral approach to teaching of origins violates the establishment clause of the United States Constitution, simply because a theory of origin may partially coincide with the tenets of some religion(s). The proof and legal authorities will show that this question must be answered with an emphatic "no." . . .

Outline of Argument

I. Act 590 Does Not Establish Religion

A. Act 590 Has A Secular Purpose   It is uniformly conceded that the State has a legitimate right to prescribe the curriculum for schools. . . . [legal citations omitted] . . . Further, the State of Arkansas' authority

*McLean v. Arkansas, U.S. District Court, Eastern District of Arkansas, Western Division, Civil Action LRC 81-322, Defendants' Outline of the Legal Issues and Proof.

over its schools is clearly recognized in Arkansas law. The Arkansas Constitution, Article 14, Section 1, grants to the State plenary authority over education. . . . The proof in this case will demonstrate that the State of Arkansas, pursuant to its conceded right to prescribe curriculum, has determined that a neutral, two-model approach to the teaching of origins is a desirable goal. In order to achieve that goal, the Arkansas Legislature passed Act 590, which requires that creation-science and evolution-science be given balanced treatment. The proof will further demonstrate that well-trained, reputable educators and scientists believe that a neutral, two-model approach to teaching of origins can be accomplished in a secular, completely non-religious manner. Further, the proof will demonstrate that the neutral, two-model approach is proper and desirable from both an educational and scientific perspective. . . .

Plaintiffs' argument on the establishment of religion relies largely upon two cases: *Epperson v. Arkansas* and *Daniel v. Waters*. Both cases are helpful background reading for the First Amendment principles which they establish. However, the facts in the instant case differ significantly from both *Epperson* and *Daniel*; accordingly, these cases carry relatively little weight in determining the outcome of this case. The decision in *Epperson* turned upon the existence of censorship of a scientific theory for religious reasons. . . . In the instant case, it is the plaintiffs who seek to censor a theory which, while perhaps controversial, is well supported by competent scientific evidence. *Epperson* did not preclude creation-science in the curriculum, it only addressed *exclusion* of Darwinian theory for religious reasons. As the evidence will show, Act 590 mandates a neutral, two-model approach to teaching of origins not for religious reasons, but for reasons related both to education and science. Thus, *Epperson* in reality supports the defendants' position in this case inasmuch as it is plaintiffs who seek to censor a scientific theory because it is contrary to their own philosophical and religious beliefs.

Daniel v. Waters similarly does not materially advance the plaintiffs' case. In *Daniel*, a Tennessee statute provided that no biology textbook which dealt with theories of origin could be used unless it specifically stated that the theory of origin was simply a theory and not representing a scientific fact. Further, the statute required that each textbook must give an equal amount of emphasis to the theories of origin besides evolution, including the Genesis account in the Bible. The textbook did not have to state, however, that the Genesis account of origins was a theory. Additionally, the statute prohibited the teaching of religious beliefs of origin which were "occult" or "satanical beliefs" (*Daniel* at 487).

The key to the holding in *Daniel* was the preferential position given the Genesis account of creation. The disclaimer which was required for most evolutionary theories was specifically not required for the Genesis story of creation. . . .

Besides finding that the Tennessee statute gave a preference to a religious theory over a scientific theory, the Court further held that preference was given to one religious concept of creation over others. . . .

Thus, the Sixth Circuit did not find that the inclusion of the biblical view of creation in a science course was invalid. It ruled only that favoritism of this view established religion.

The decision in *Daniel* clearly suggests that a state legislature may require diversity in presentation of theories of origin wherever the subject of origins is to be presented. This is exactly what Act 590 is designed to do and it accomplishes its result without the inclusion of any religious writings or teachings.

At this juncture, it appears that plaintiffs seek to construct a complete wall of separation between church and state which the Supreme Court has never recognized. On the contrary, the Court has clearly stated that the establishment clause does "not call for a total separation of church and state," and that the requirement, "far from being in fact a 'wall' is a blurred, indistinct, and variable barrier . . ." (*Lemon v. Kurtzman* at 614).

B. The Principal Effect of Act 590 Neither Advances Nor Inhibits Religion If plaintiffs' brief is any preview of their case, plaintiffs hope to prove that Act 590 advances religion, by relying upon writings with religious connotations which are in fact specifically prohibited by Act 590. . . . Merely because a publication uses the term "creation-science" does not mean that the publication is in conformity with Act 590. Such twisted logic need not long detain the Court.

Beyond this make-weight argument, plaintiffs rely principally upon the concept that creation-science requires a creator, and that a creator is inherently a religious belief. . . . Plaintiffs believe that once the word "creator" is used in teaching about origin, the Establishment Clause has been violated. Such a simplistic view overlooks not only the legal authorities but also scientific thought, as well. For example, reference to "God" in the pledge to the United States flag or to "In God is our trust' in the National Anthem in public school classrooms has been held not to violate the establishment clause (*Engel v. Vitale* at 435, *Aronow v. United States* at 243). Even the seminal volume on evolution written by Charles Darwin mentions a "creator" several times and it ends with the statement that "[T]here is grandeur in this view of life . . . having been originally breathed by the Creator into a few forms" (Charles Darwin, *The Origin of Species*, p. 759). Thus, if plaintiffs' theory is correct, then Darwin's own writings must be excluded from the science classrooms. . . .

Rather than advancing religion, Act 590 advances both scientific inquiry and academic freedom. Scientific inquiry is advanced by providing students with an alternative scientific theory to evolution-science. The proof herein will show that much of the scientific evidence concerning origins supports the theory of creation-science, and that other scientific evidence does not support evolution-science. As such, creation-science is equally as scientific as evolution-science.

Further, Act 590 has a primary effect of furthering academic freedom in that a controversial scientific theory should not be squelched or censored by any self-appointed guardians of what is "science." See the discussion of academic freedom *infra*.

C. Act 590 Does Not Create Excessive Entanglement Between the State and Religion   While it is clear that "excessive entanglement" between religion and government will violate a statute under the Establishment Clause (see *Lemon v. Kurtzman* at 612–613), defining what is "excessive entanglement" does not admit of any easy answer. . . .

. . . [T]here will be, in fact, no entanglement of the state and religion inasmuch as creation-science is a scientific inquiry. The mere fact that the scientific evidence for creation-science may coincide with the tenets or beliefs of some religions clearly does not result in the establishment of those religions. In *Harris v. McRae*, the Supreme Court summarily rejected the argument that limiting Medicaid funds for abortion violated the establishment clause because such a requirement incorporated into the law the doctrines of the Roman Catholic Church. The Court [in *Harris v. McRae*] reasoned that:

Although neither a State nor Federal Government can constitutionally 'pass laws which aid one religion, aid all religions, or prefer one religion over another' (*Everson v. Board of Education* at 15), it does not follow that a statute violates the establishment clause because it 'happens to coincide or harmonize with the tenets of some or all religions' (*McGowan v. Maryland* at 442).

Stripped of its embellishment, this is precisely what plaintiffs are arguing in this case. . . . [B]oth criminal laws (e.g., theft, murder) and civil laws (e.g., contractual obligations, marriage contracts) are often rooted in religious teachings or are consistent with the tenets of many religions. . . . However, assuming Act 590 may be consistent with some individuals' religions beliefs, both *Harris* and *McGowan* authoritatively establish that mere consistency of laws and religious doctrines does not result in a violation of the establishment clause.

II. Act 590 Does Not Unconstitutionally Abridge Academic Freedom Rights

Although the United States Supreme Court has recognized that the freedom of both teachers and students to "inquire, to study and to evaluate, to gain new maturity and understanding . . . is essential" (*Sweezy v. New Hampshire* at 250), the Court has stopped far short of placing "academic freedom" among the fundamental rights guaranteed by the Constitution. Instead, the Court has adopted a balancing test whereby the State's interest in assuring that courts and programs are properly presented is weighed against any restrictions and limitations placed upon the rights of teachers and pupils (*Epperson v. Arkansas*).

In the instant case, plaintiffs argue that Act 590 impinges upon the academic freedom of teachers to exercise their professional judgment in selecting what should be presented in the classroom. . . . However, in their zeal to limit the right of teachers to teach a controversial scientific theory, i.e., creation-science, plaintiffs have overlooked the recognized fact [that] the State of Arkansas has the power to establish and prescribe school curriculum. Accordingly, plaintiffs have failed to properly identify and apply the countervailing weight in the balancing test for academic freedom.

The right of a teacher to determine the scope of classroom instruction is limited by the right of the State to prescribe curriculum. While the State cannot adopt a course of instruction which would aid or oppose any religion, it is free to adopt programs and practices for use in the public schools which ensure religious neutrality. The evidence will show that Act 590 is not an attempt by the State to advance or inhibit any religion. In fact, §2 of the Act specifically provides that "[t]reatment of either evolution-science or creation-science shall be limited to scientific evidences for each model and inferences from those scientific evidences, *and must not include any religious instruction or references to religious writings*" (emphasis added).

In *Epperson v. Arkansas*, a case heavily relied upon by plaintiffs, the Supreme Court noted that the Arkansas law prohibiting the teaching of evolution in the public schools was designed to suppress a theory of origins, contrary to the beliefs of many citizens. Ironically, plaintiffs in the present case, like the defendants in *Epperson*, have taken up the cause of academic censorship and joined forces with those who would suppress the teaching of a scientific theory of origins simply because they do not personally agree with it and because it is incompatible with their own religious or philosophical beliefs. Academic freedom demands that teachers be permitted to instruct their students in all known theories of origins. The relative strengths of each theory should also be discussed in order to preserve academic integrity. The proof will show that there is scientific evidence to support the creation-science view of origins. Accordingly, teachers and students should be free to discuss such evidence without threat of reprisal from those who hold contrary views. No group, regardless of its unselfish concern for the welfare of others, should take away the right of teachers to teach and students to learn about even the most controversial issue. . . . In the case at bar, creation-science should not be suppressed from the marketplace of ideas. Rather, its merits should be discussed and criticized by the academic community inside, as well as outside, of the classroom. . . .

Act 590 does not abridge academic freedom rights by censoring a theory of origins. On the contrary, the evidence will show that, but for the protection offered by the Act, some teachers would not teach creation-science. Moreover, lest plaintiffs seek to confuse the issues still further, Act 590 does not prohibit critical inquiry by teachers of either model of origins. The Act simply guarantees that students will be exposed to the two models of origins, if either model is discussed in the classroom.

Plaintiffs complain that in passing Act 590, the Arkansas Legislature gave special recognition to the place of creation-science in the classroom. . . . However, the decision of the Legislature is neither inappropriate nor improper. . . . In the present case, the people of the State of Arkansas, acting through their elected representatives, have chosen to expose students in their public schools to a neutral, two-model approach of origins. This decision is a legitimate exercise of the state's power to set curriculum in its public schools. This Court should not superimpose its judgment for that of the people simply because plaintiffs have raised the banner of academic freedom. . . . Act 590 ensures academic freedom to teachers and students to explore a controversial theory of origins. The evidence will show that the primary effect of the enactment will be to promote a full and open discussion of origins, without aiding or opposing any religion. As such, the academic freedom of teachers and students is fully protected.

III. Act 590 Is Not Unconstitutionally Vague

Act 590 sets forth with specificity a definition of creation-science. Section 4(a) of Act 590 defines creation-science. . . . Further, the act specifically limits treatment of both evolution-science and creation-science to the scientific evidences for each model and explicitly prohibits the use of any religious instruction or references to religious writings.

The term "balanced treatment" is enlarged upon in Sections 1 and 5 of the Act. . . .

Despite these specific instructions in the Act, the plaintiffs nonetheless allege that Act 590 is unconstitutionally vague. At the outset, it should be noted that plaintiffs' arguments place them on the horns of a dilemma. On the one hand they argue that Act 590 unconstitutionally abridges academic freedom in that the legislature has prescribed the manner in which a particular subject must be taught. Plaintiffs then belie their own position on academic freedom by alleging that the Act is unconstitutionally vague because teachers do not know what they are to teach. Thus, the inconsistency of these two positions simply reveals that both are transparently makeweight attempts to fashion an argument of constitutional dimensions where none exist.

With regard to the vagueness argument, the linchpin of plaintiffs' case is that a non-probationary teacher's contract may not be terminated nor may it be nonrenewed except for cause which is "not arbitrary, capricious or discriminatory." . . . This argument is an attempt by plaintiffs to bootstrap their vagueness argument on the basis of a State statute. . . .

In measuring a vagueness argument, a noncriminal statute is not unconstitutionally vague where the statute is set out in terms that the ordinary person exercising ordinary common sense can sufficiently understand and comply with the statute. . . . [legal citations omitted] . . . In the present case, this standard of an "ordinary person exercising ordinary common sense" must be read in light of the fact that Act 590 is to be implemented by professionally trained educators. The proof will show that a professional educator can give "balanced treatment" within the meaning of Act 590. Perhaps more importantly, however, by requiring "balanced treatment," rather than "equal time" or some other more exact standard, Act 590 leaves it to the professional judgment, training, and discretion of the teacher in the classroom on how the treatment of both evolution-science and creation-science should be balanced.

Two additional points must be made concerning the vagueness argument. First, it is entirely speculative to argue that the statute is unconstitutionally vague inasmuch as it is not yet in effect. . . . [T]he issue in this case is one of whether the statute is valid on its face. If found facially valid, the plaintiffs cannot forward an argument that it *might* be applied in an unconstitutional manner at some uncertain future date. A plaintiff cannot be heard to attack a constitutional statute on the grounds that it might someday be applied to him in an unconstitutional manner. . . . [legal citations omitted] . . . Second, assuming for the purposes of argument that plaintiffs' vagueness allegation has some merit, this Court should abstain from reaching that question until such time that a state court can interpret the statute and perhaps obviate the federal constitutional question. (See, for example, *Daniel v. Waters.*)

Conclusion
For the foregoing reasons and based on the evidence which will be presented at trial, defendants will ask that this Court declare Act 590 constitutional under the First and Fourteenth Amendments to the United States Constitution.

EXCERPTS FROM DEFENDANTS' TRIAL BRIEF*

[*Editor's note.* In its Trial Brief, the State of Arkansas expanded on its argument that the "principal effect of Act 590 neither advances nor inhibits religion." This section, beginning at page 14 of the brief, contains the discussion referred to in Langdon Gilkey's essay in this volume, of the necessary characteristics of a "creator" and a "god."]

. . . Plaintiffs argue that by adopting a neutral approach to teaching of origins, the state has violated the establishment clause because creation-science may partially coincide with the tenets of some religions. A similar argument was recently rejected by the Eighth Circuit in holding that a neutral policy of a state university which would permit religious groups to use university buildings for religious services would not establish religion (*Chess v. Widmar*). Indeed, if the coincidence of creation-

*Excerpted from *McLean v. Arkansas*, U.S. District Court, Eastern District of Arkansas, Western Division, Civil Action LRC 81-322, Defendants' Trial Brief.

science to the tenets of some religions violates the First Amendment, then the presentation of evolution-science must necessarily also violate the Constitution. The proof will demonstrate that evolution-science is as equally consistent with the tenets of many religions as creation-science may be with the others.

Further, the proof will show that there is nothing inherently religious about the terms "creator" or "creation," as used in the context of Act 590. Act 590 is concerned with a non-religious conception of "creation" and "creator," not the religious concepts dealt with in the Bible or religious writings. Similarly, the proof will show that there is a valid scientific principle concerning evolution called "teleology." Expert witnesses (including plaintiffs' experts) will testify that there can be both religious and non-religious theories of teleology, depending on how the term is used. Religious teleology is based on the workings of a religious being; non-religious teleology is a valid scientific principle devoid of any religious beliefs.

Assuming *arguendo* that a "creator" and "creation" are consistent with some religions, this does not make these inherently religious. The entity which caused the creation hypothesized in creation-science is far, far away from any conception of a god or deity. All that creation-science requires is that the entity which caused creation have power, intelligence, and a sense of design. There are no attributes of the personality generally associated with a deity, nor is there necessarily present in the creator any love, compassion, sense of justice, or concern for any individuals. Indeed, under creation-science as defined in Act 590, there is no requirement that the entity which caused creation still be in existence. . . .

. . . The mere fact that the scientific evidence for creation-science may coincide with the tenets or beliefs of some religions clearly does not result in the establishment of those religions (See *Harris v. McRae*). . . .

The study of creation-science will no more commit the State to the religious beliefs held by some individuals associated with the creation-science "movement" (if such exists) than the study of a particular theory of economics (e.g., Keynesian, supply-side, etc.) or political philosophy (e.g., marxism, socialism, democracy) would signal that the state endorsed the theories or beliefs personally held by the adherents of these theories (See *Chess v. Widmar* at 1317). . . .

. . . Rather than advancing religion, Act 590 advances both scientific inquiry and academic freedom. Scientific inquiry is advanced by providing students with an alternative scientific theory to evolution-science.

The proof will show that many competent scientists believe that the scientific data on origins best support creation-science.

Act 590 has a primary effect of furthering academic freedom in that a controversial scientific theory should not be squelched or censored based on one small segment of society's political, philosophical, or religious opposition to the theory. The plaintiffs' witnesses are in many cases opposed to creation-science on grounds of politics, philosophy, and religion. . . .

[*Editor's note.* The Trial Brief for the defense also contained a section (beginning at page 23) not included in the "Preliminary Outline," in which the State argues that "Act 590 restores substantial neutrality in the teaching of origins."]

. . . Underlying the familiar three-pronged test for determining whether the establishment clause has been violated is one fundamental theme: the government must remain neutral in matters of religion.

Government in our democracy, state and national, must be neutral in matters of religious theory, doctrine, and practice. It may not be hostile to any religion or to the advocacy of non-religion; and it may not aid, foster, or promote one religion or religious theory against another or even against the militant opposition. The First Amendment mandates governmental neutrality between religion and religion and between religion and non-religion (*Epperson v. Arkansas* at 103–104).

The proof herein will demonstrate that evolution-science is consistent with the tenets of many religions, just as creation-science is consistent with others. Further, evolution is a fundamental tenet of some theistic and nontheistic religions. . . .

The Supreme Court has recognized that there exist religions in this country which do not include traditional theistic beliefs.

Among religions in this country which do not teach what would generally be considered a belief in the existence of God are Buddhism, Taoism, Ethical Culture, Secular Humanism, and others (*Torcaso v. Watkins* at 495).

These religious belief systems are, therefore, subject to First Amendment protection and prohibition. Evolution is an important doctrine in each of these religions. . . .

. . . [T]here is no requirement of a deity or supreme being in order to qualify as a religion under the First Amendment. Therefore, in light of this fact, let us turn to an examination of evolution-science and religion.

Evolution-science is consistent with many religious beliefs and inconsistent with others. In fact, the theological liberalists, Secular Humanists and Religious Humanists heavily immesh their religious beliefs with principles of evolution-science (See Whitehead and Conlan, "The Establishment of the Religion of Secular Humanism and Its First Amendment Implications," 10 *Texas Tech Law Review* 1, 46–54, 1978, and Wendell Bird, "Freedom from Establishment and Unneutrality in Public School Instruction and Religious School Regulation" 2 *Harvard Journal of Law and Public Policy* 125, 198–204, 1979). Not only is evolution merely consistent with some religions, in others it is essential.

In fact, "Secular Humanism as a religion is incomprehensible without the evolutionary hypothesis. The evolutionary hypothesis is one tenet, if extracted, that will disembowel Secular Humanism" (Whitehead and Conlan, *Idem* at 54). . . .

[*Editor's note.* In the section of the Brief arguing that Act 590 does *not* unconstitutionally abridge academic freedom, the State included this paragraph pertinent to the case as a whole:]

. . . There is no school system which can reasonably expect to teach all knowledge. Choices must necessarily be made. As one court has stated, "The whole range of knowledge and ideas cannot be taught in the limited time available in public schools. . . . The authorities must choose which portions of the world's knowledge will be included in the curriculum's programs and courses and which portions will be left for grasping from other sources such as the family, peers, or other institutions" (*Mercer v. Michigan State Board of Education*). If anything is clear in educational policy and curriculum today, it is beyond doubt that no matter what curriculum decision is made, some group will be offended (*President's Council District 25 v. Community School Board District 25*). Offense may often, as in the present case, lead to litigation, but that does not mean that curriculum is best served by litigation. Indeed, the contrary is more likely true according to the Second Circuit: "Academic freedom is scarcely fostered by the intrusion of federal jurists making curriculum or library choices for the community of scholars" (*Idem*). . . .

Cases Cited

Aronow v. United States, 432 F.2d 242 (9th Circuit 1970).

Brandon v. Board of Education, 487 F. Supp 1219 (N.D.N.Y.), *affirmed*, 635 F.2d 971 (2d Circuit 1980).

Chess v. Widmar, 635 F.2d 1310 (8th Circuit 1980), *certiorari granted*, 480 U.S. 907, 101 S. Ct. 1345, 67 L.Ed. 2d 332 (1981).

Connally v. General Construction Company, 269 U.S. 385 (1926).

Daniel v. Waters, 515 F.2d 485 (6th Circuit 1975).

Engel v. Vitale, 370 U.S. 421 (1962).

Epperson v. Arkansas, 393 U.S. 97 (1968).

Everson v. Board of Education, 330 U.S. 1 (1947).

Gravned v. City of Rockford, 408 U.S. 104 (1972).

Hall v. Bradshaw, 630 F.2d 1018 (4th Circuit 1980).

Harris v. McRae, 448 U.S. 297 (1980).

Kevishian v. Board of Regents, 385 U.S. 589 (1967).

Lemon v. Kurtzman, 403 U.S. 602 (1971).

McGowan v. Maryland, 366 U.S. 420 (1961).

Malnak v. Yogi, 440 F. Supp. 1284 (D.N.J. 1977), *affirmed per curiam*, 592 F.2d 197 (3d Circuit 1979).

Mercer v. Michigan State Board of Education, 379 F. Supp. 580 (E.D. Mich. 1974), *affirmed mem.*, 419 U.S. 1081 (1975).

Moore v. Gaston County Board of Education, 357 F. Supp. 1037 (W.D.N.C. 1973).

Pico v. Board of Education, 638 F.2d 404 (2d Circuit 1980).

President's Council District 25 v. Community School Board District 25, 457 F.2d 289 (1972).

Stone v. Graham, 449 U.S. 39 (1980).

Sweezy v. New Hampshire, 354 U.S. 234 (1957).

Torcaso v. Watkins, 367 U.S. 488 (1961).

Webb v. Lake Mills Community School District, 344 F. Supp. 791 (N.D. Iowa 1972).

McLean v. Arkansas

Opinion of William R.
Overton, U.S. District
Judge, Eastern District of
Arkansas, Western Division
(Dated 5 January 1982)

I

There is no controversy over the legal standards under which the Establishment Clause portion of this case must be judged. The Supreme Court has on a number of occasions expounded on the meaning of the clause, and the pronouncements are clear. Often the issue has arisen in the context of public education, as it has here. In *Everson v. Board of Education*, Justice Black stated:

The 'establishment of religion' clause of the First Amendment means at least this: Neither a state nor the Federal Government can set up a church. Neither can pass laws which aid one religion, aid all religions, or prefer one religion over another. Neither can force nor influence a person to go to or to remain away from church against his will or force him to profess a belief or disbelief in any religion. No person can be punished for entertaining or professing religious beliefs or disbeliefs, for church attendance or non-attendance. No tax, large or small, can be levied to support any religious activities or institutions, whatever they may be called, or whatever form they may adopt to teach or practice religion. Neither a state nor the Federal Government can, openly or secretly, participate in the affairs of any religious organizations or groups and *vice versa*. In the words of Jefferson, the clause . . . was intended to erect 'a wall of separation between church and State.' (*Everson* at 15–16.)

The Establishment Clause thus enshrines two central values: voluntarism and pluralism. And it is in the area of the public schools that these values must be guarded most vigilantly.

Designed to serve as perhaps the most powerful agency for promoting cohesion among a heterogeneous democratic people, the public school must keep scrupulously free from entanglement in the strife of sects. The preservation of the community from divisive conflicts, of Government from irreconcilable pressures by religious groups, of religion from censorship and coercion however subtly exercised, requires strict confinement of the State to instruction other than religious, leaving to the individual's church and home, indoctrination in the faith of his choice.[*McCollum v. Board of Education* at 216–217 (Opinion of Justice Frankfurter, joined by Justices Jackson, Burton, and Rutledge).]

The specific formulation of the establishment prohibition has been refined over the years, but its meaning has not varied from the principles articulated by Justice Black in *Everson.* In *Abbington School District v. Schempp*, Justice Clark stated that "to withstand the strictures of the Establishment Clause there must be a secular legislative purpose and a primary effect that neither advances nor inhibits religion." The Court found it quite clear that the First Amendment does not permit a state to require the daily reading of the Bible in public schools, for "[s]urely the place of the Bible as an instrument of religion cannot be gainsaid" (*Idem* at 224). Similarly, in *Engel v. Vitale*, the Court held that the First Amendment prohibited the New York Board of Regents from requiring the daily recitation of a certain prayer in the schools. With characteristic succinctness, Justice Black wrote, "Under [the First] Amendment's prohibition against governmental establishment of religion, as reinforced by the provisions of the Fourteenth Amendment, government in this country, be it state or federal, is without power to prescribe by law any particular form of prayer which is to be used as an official prayer in carrying on any program of governmentally sponsored religious activity" (*Idem* at 430). Black also identified the objective at which the Establishment Clause was aimed: "Its first and most immediate purpose rested on the belief that a union of government and religion tends to destroy government and to degrade religion" (*Idem* at 431).

Most recently, the Supreme Court has held that the clause prohibits a state from requiring the posting of the Ten Commandments in public school classrooms for the same reasons that officially imposed daily Bible reading is prohibited (*Stone v. Graham*). The opinion in *Stone* relies on the most recent formulation of the Establishment Clause test, that of *Lemon v. Kurtzman*:

First, the statute must have a secular legislative purpose; second, its principal or primary effect must be one that neither advances nor inhibits religion. . . ; finally, the statute must not foster 'an excessive government entanglement with religion' (*Stone* at 40).

It is under this three-part test that the evidence in this case must be judged. Failure on any of these grounds is fatal to the enactment.

II

The religious movement known as Fundamentalism began in nineteenth-century America as part of evangelical Protestantism's response to social changes, new religious thought, and Darwinism. Fundamentalists viewed these developments as attacks on the Bible and as responsible for a decline in traditional values.

The various manifestations of Fundamentalism have had a number of common characteristics,[4] [*Editor's note.* Footnotes follow the numbering scheme as given in the original text of the court opinion. Notes 1, 2, and 3 were cited in the "Introduction" reviewing the Bill's history, which has been omitted.] but a central premise has always been a literal interpretation of the Bible and a belief in the inerrancy of the Scriptures. Following World War I, there was again a perceived decline in traditional morality, and Fundamentalism focused on evolution as responsible for the decline. One aspect of their efforts, particularly in the South, was the promotion of statutes prohibiting the teaching of evolution in public schools. In Arkansas, this resulted in the adoption of Initiated Act 1 of 1929.[5]

Between the 1920s and early 1960s, anti-evolutionary sentiment had a subtle but pervasive influence on the teaching of biology in public schools. Generally, textbooks avoided the topic of evolution and did not mention the name of Darwin. Following the launch of the Sputnik satellite by the Soviet Union in 1957, the National Science Foundation

4. The authorities differ as to generalizations which may be made about Fundamentalism. For example, Dr. Geisler testified to the widely held view that there are five beliefs characteristic of all Fundamentalist movements, in addition, of course, to the inerrancy of Scripture: (1) belief in the virgin birth of Christ, (2) the belief in the deity of Christ, (3) belief in the substitutional atonement of Christ, (4) belief in the second coming of Christ, and (5) belief in the physical resurrection of all departed souls. Dr. Marsden, however, testified that this generalization, which had been common in religious scholarship, is now thought to be historical error. There is no doubt, however, that all Fundamentalists take the Scriptures as inerrant and probably most take them as literally true.
5. Initiated Act 1 of 1929, Arkansas Statutes Ann. § 80–1627 *et seq.*, which prohibited the teaching of evolution in Arkansas schools is discussed *infra* at text accompanying note 15 (below).

funded several programs designed to modernize the teaching of science in the nation's schools. The Biological Sciences Curriculum Study (BSCS), a nonprofit organization, was among those receiving grants for curriculum study and revision. Working with scientists and teachers, BSCS developed a series of biology texts which, although emphasizing different aspects of biology, incorporated the theory of evolution as a major theme. The success of the BSCS effort is shown by the fact that fifty percent of American school children currently use BSCS books directly and the curriculum is incorporated indirectly in virtually all biology texts (Testimony of Mayer, Nelkin; Plaintiffs' Exhibit 1).[6]

In the early 1960s, there was again a resurgence of concern among Fundamentalists about the loss of traditional values and a fear of growing secularism in society. The Fundamentalist movement became more active and has steadily grown in numbers and political influence. There is an emphasis among current Fundamentalists on the literal interpretation of the Bible and the Book of Genesis as the sole source of knowledge about origins.

The term "scientific creationism" first gained currency around 1965 following publication of *The Genesis Flood* in 1961 by Whitcomb and Morris. There is undoubtedly some connection between the appearance of the BSCS texts emphasizing evolutionary thought and efforts by Fundamentalists to attack the theory (Testimony of Mayer).

In the 1960s and early 1970s, several Fundamentalist organizations were formed to promote the idea that the Book of Genesis was supported by scientific data. The terms "creation science" and "scientific creationism" have been adopted by these Fundamentalists as descriptive of their study of creation and the origins of man. Perhaps the leading creationist organization is the Institute for Creation Research (ICR), which is affiliated with the Christian Heritage College and supported by the Scott Memorial Baptist Church in San Diego, California. The ICR, through the Creation-Life Publishing Company, is the leading publisher of creation science material. Other creation science organizations include the Creation Science Research Center (CSRC) of San Diego and the Bible Science Association of Minneapolis, Minnesota. In 1963, the Creation Research Society (CRS) was formed from a schism in the American Scientific Affiliation (ASA). It is an organization

6. References to documentary exhibits are by the name of the author and the [trial or pre-trial] exhibit number.

of literal Fundamentalists[7] who have the equivalent of a master's degree in some recognized area of science. A purpose of the organization is "to reach all people with the vital message of the scientific and historic truth about creation" (Nelkin, *The Science Textbook Controversies and the Politics of Equal Time*, p. 66). Similarly, the CSRC was formed in 1970 from a split in the CRS. Its aim has been "to reach the 63 million children of the United States with the scientific teaching of Biblical creationism" (*Idem* at 69).

Among creationist writers who are recognized as authorities in the field by other creationists are Henry M. Morris, Duane Gish, G. E. Parker, Harold S. Slusher, Richard B. Bliss, John W. Moore, Martin E. Clark, W. L. Wysong, Robert E. Kofahl, and Kelly L. Segraves. Morris is Director of ICR, Gish is Associate Director, and Segraves is associated with CSRC.

Creationists view evolution as a source of society's ills, and the writings of Morris and Clark are typical expressions of that view.

Evolution is thus not only anti-Biblical and anti-Christian, but it is utterly unscientific and impossible as well. But it has served effectively as the pseudo-scientific basis of atheism, agnosticism, socialism, fascism, and numerous other false and dangerous philosophies over the past century. [Morris and Clark, *The Bible Has the Answer* (Plaintiffs' Exhibit 31 and Plaintiffs' Pre-trial Exhibit 89).][8]

7. Applicants for membership in the CRS must subscribe to the following statement of belief: "(1) The Bible is the written Word of God, and because we believe it to be inspired thruout (sic), all of its assertions are historically and scientifically true in all of the original autographs. To the student of nature, this means that the account of origins in Genesis is a factual presentation of simple historical truths. (2) All basic types of living things, including man, were made by direct creative acts of God during Creation Week as described in Genesis. Whatever biological changes have occurred since Creation have accomplished only changes within the original created kinds. (3) The great Flood described in Genesis, commonly referred to as the Noachian Deluge, was an historical event, worldwide in its extent and effect. (4) Finally, we are an organization of Christian men of science, who accept Jesus Christ as our Lord and Savior. The account of the special creation of Adam and Eve as one man and one woman, and their subsequent Fall into sin, is the basis for our belief in the necessity of a Savior for all mankind. Therefore, salvation can come only thru (sic) accepting Jesus Christ as our Savior." (Plaintiffs' Exhibit 115)

8. Because of the voluminous nature of the documentary exhibits, the parties were directed by pre-trial order to submit their proposed exhibits for the Court's convenience prior to trial. The numbers assigned to the pre-trial submissions do not correspond with those assigned to the same documents at trial and, in some instances, the pre-trial submissions are more complete.

Creationists have adopted the view of Fundamentalists generally that there are only two positions with respect to the origins of the earth and life: belief in the inerrancy of the Genesis story of creation and of a worldwide flood as fact, or belief in what they call evolution.

Henry Morris has stated, "It is impossible to devise a legitimate means of harmonizing the Bible with evolution" [Morris, "Evolution and the Bible," *ICR Impact Series,* Number 5 (undated, unpaged), quoted in Mayer, Plaintiffs' Exhibit 8, p. 3]. This dualistic approach to the subject of origins permeates the creationist literature.

The creationist organizations consider the introduction of creation science into the public schools [to be] part of their ministry. The ICR has published at least two pamphlets[9] containing suggested methods for convincing school boards, administrators and teachers that creationism should be taught in public schools. The ICR has urged its proponents to encourage school officials to voluntarily add creationism to the curriculum.[10]

Citizens For Fairness In Education is an organization based in Anderson, South Carolina, formed by Paul Ellwanger, a respiratory therapist who is trained in neither law nor science. Mr. Ellwanger is of the opinion that evolution is the forerunner of many social ills, including Nazism, racism and abortion (Ellwanger Deposition, pp. 32–34). About 1977, Ellwanger collected several proposed legislative acts with the idea of preparing a model state act requiring the teaching of creationism as science in opposition to evolution. One of the proposals he collected was prepared by Wendell Bird, who is now a staff attorney for ICR.[11]

---

9. Plaintiffs' Exhibit 130, Morris, *Introducing Scientific Creationism into the Public Schools* (1975); and Bird, "Resolution for Balanced Presentation of Evolution and Scientific Creationism," *ICR Impact Series* No. 71, Appendix 14 to Plaintiffs' Pretrial Brief.

10. The creationists often show candor in their proselytization. Henry Morris has stated, "Even if a favorable statute or court decision is obtained, it will probably be declared unconstitutional, especially if the legislation or injunction refers to the Bible account of creation." In the same vein he notes, "The only effective way to get creationism taught properly is to have it taught by teachers who are both willing and able to do it. Since most teachers now are neither willing nor able, they must first be persuaded and instructed themselves." [Plaintiffs' Exhibit 130, Morris, *Introducing Scientific Creationism into the Public Schools* (1975) (unpaged).]

11. Mr. Bird sought to participate in this litigation by representing a number of individuals who wanted to intervene as defendants. The application for intervention was denied by this Court. . . .

From these various proposals, Ellwanger prepared a "model act" which calls for "balanced treatment" of "scientific creationism" and "evolution" in public schools. He circulated the proposed act to various people and organizations around the country.

Mr. Ellwanger's views on the nature of creation science are entitled to some weight since he personally drafted the model act which became Act 590. His evidentiary deposition with exhibits and unnumbered attachments (produced in response to a subpoena *duces tecum*) speaks to both the intent of the Act and the scientific merits of creation science. Mr. Ellwanger does not believe creation science is a science. In a letter to Pastor Robert E. Hays he states, "While neither evolution nor creation can qualify as a scientific theory, and since it is virtually impossible at this point to educate the whole world that evolution is not a true scientific theory, we have freely used these terms—the evolution theory and the theory of scientific creationism—in the bill's text" (Unnumbered attachment to Ellwanger Deposition, p. 2). He further states in a letter to Mr. Tom Bethell, "As we examine evolution (remember, we're not making scientific claims for creation, but we are challenging evolution's claim to be scientific) . . ." (Unnumbered attachment to Ellwanger Deposition, p. 1).

Ellwanger's correspondence on the subject shows an awareness that Act 590 is a religious crusade, coupled with a desire to conceal this fact. In a letter to State Senator Bill Keith of Louisiana, he says, "I view this whole battle as one between God and anti-God forces, though I know there are a large number of evolutionists who believe in God." And further, ". . . it behooves Satan to do all he can to thwart our efforts and confuse the issue at every turn." Yet Ellwanger suggests to Senator Keith, "If you have a clear choice between having grassroots leaders of this statewide bill promotion effort to be ministerial or non-ministerial, be sure to opt for the non-ministerial. It does the bill effort no good to have ministers out there in the public forum and the adversary will surely pick at this point. . . . Ministerial persons can accomplish a tremendous amount of work from behind the scenes, encouraging their congregations to take the organizational and P.R. initiatives. And they can lead their churches in storming Heaven with prayers for help against so tenacious an adversary" (Unnumbered attachment to Ellwanger Deposition, p. 1).

Ellwanger shows a remarkable degree of political candor, if not finesse, in a letter to State Senator Joseph Carlucci of Florida:

2. It would be very wise, if not actually essential, that all of us who are engaged in this legislative effort be careful not to present our position and our work in a religious framework. For example, in written communications that might somehow be shared with those other persons whom we may be trying to convince, it would be well to exclude our own personal testimony and/or witness for Christ, but rather, if we are so moved, to give that testimony on a separate attached note. (Unnumbered attachment to Ellwanger Deposition, p. 1.)

The same tenor is reflected in a letter by Ellwanger to Mary Ann Miller, a member of FLAG (Family, Life, America under God) who lobbied the Arkansas Legislature in favor of Act 590:

... we'd like to suggest that you and your co-workers be very cautious about mixing creation-science with creation-religion. ... Please urge your co-workers not to allow themselves to get sucked into the 'religion' trap of mixing the two together, for such mixing does incalculable harm to the legislative thrust. It could even bring public opinion to bear adversely upon the higher courts that will eventually have to pass judgment on the constitutionality of this new law. [Exhibit 1 to Miller Deposition.]

Perhaps most interesting, however, is Mr. Ellwanger's testimony in his deposition as to his strategy for having the model act implemented:

Q. You're trying to play on other people's religious motives.

A. I'm trying to play on their emotions, love, hate, their likes, dislikes, because I don't know any other way to involve, to get humans to become involved in human endeavors. I see emotions as being a healthy and legitimate means of getting people's feelings into action, and ... I believe that the predominance of population in America that represents the greatest potential for taking some kind of action in this area is a Christian community. I see the Jewish community as far less potential in taking action ... but I've seen a lot of interest among Christians and I feel, why not exploit that to get the bill going if that's what it takes. (Ellwanger Deposition, pp. 146–147.)

Mr. Ellwanger's ultimate purpose is revealed in the closing of his letter to Mr. Tom Bethell: "Perhaps all this is old hat to you, Tom, and if so, I'd appreciate your telling me so and perhaps where you've heard it before—the idea of killing evolution instead of playing these debating games that we've been playing for nigh over a decade already" (Unnumbered attachment to Ellwanger Deposition, p. 3).

It was out of this milieu that Act 590 emerged. The Reverend W. A. Blount, a Biblical literalist who is pastor of a church in the Little Rock area and was, in February, 1981, chairman of the Greater Little

Rock Evangelical Fellowship, was among those who received a copy of the model act from Ellwanger.[12]

At Reverend Blount's request, the Evangelical Fellowship unanimously adopted a resolution to seek introduction of Ellwanger's act in the Arkansas Legislature. A committee composed of two ministers, Curtis Thomas and W. A. Young, was appointed to implement the resolution. Thomas obtained from Ellwanger a revised copy of the model act which he transmitted to Carl Hunt, a business associate of Senator James L. Holsted, with the request that Hunt prevail upon Holsted to introduce the Act.

Holsted, a self-described "born again" Christian Fundamentalist, introduced the Act in the Arkansas Senate. He did not consult the State Department of Education, scientists, science educators, or the Arkansas Attorney General.[13] The Act was not referred to any Senate committee for hearing and was passed after only a few minutes' discussion on the Senate floor. In the House of Representatives, the bill was referred to the Education Committee which conducted a perfunctory fifteen-minute hearing. No scientist testified at the hearing, nor was any representative from the State Department of Education called to testify.

Ellwanger's model act was enacted into law in Arkansas as Act 590 without amendment or modification other than minor typographical changes. The legislative "findings of fact" in Ellwanger's act and Act 590 are identical, although no meaningful fact-finding process was employed by the General Assembly.

Ellwanger's efforts in preparation of the model act and campaign for its adoption in the states were motivated by his opposition to the theory of evolution and his desire to see the Biblical version of creation taught in the public schools. There is no evidence that the pastors, Blount, Thomas, Young, or the Greater Little Rock Evangelical Fellowship were motivated by anything other than their religious convictions when proposing its adoption or during their lobbying efforts in its behalf. Senator Holsted's sponsorship and lobbying efforts in behalf of the Act

12. The model act had been revised to insert "creation science" in lieu of creationism because Ellwanger had the impression people thought creationism was too religious a term (Ellwanger Deposition, p. 79).
13. The original model act had been introduced in the South Carolina Legislature, but had died without action after the South Carolina Attorney General had opined that the act was unconstitutional.

were motivated solely by his religious beliefs and desire to see the Biblical version of creation taught in the public schools.[14]

The State of Arkansas, like a number of states whose citizens have relatively homogeneous religious beliefs, has a long history of official opposition to evolution which is motivated by adherence to Fundamentalist beliefs in the inerrancy of the Book of Genesis. This history is documented in Justice Fortas' opinion in *Epperson v. Arkansas*, which struck down Initiated Act 1 of 1929, Ark. Stat. Ann. §§80-1627–1628, prohibiting the teaching of the theory of evolution. To this same tradition may be attributed Initiated Act 1 of 1930, Ark. Stat. Ann. §80-1606 (Repealed 1980), requiring "the reverent daily reading of a portion of the English Bible" in every public school classroom in the State.[15]

It is true, as defendants argue, that courts should look to legislative statements of a statute's purpose in Establishment Clause cases and accord such pronouncements great deference (See, e.g., *Committee for Public Education & Religious Liberty v. Nyquist* at 773 and *McGowan v. Maryland* at 445). Defendants also correctly state the principle that remarks by the sponsor or author of a bill are not considered controlling in analyzing legislative intent (See, e.g., *United States v. Emmons* and *Chrysler Corporation v. Brown*).

Courts are not bound, however, by legislative statements of purpose or legislative disclaimers (*Stone v. Graham, Abbington School District v. Schempp*). In determining the legislative purpose of a statute, courts may consider evidence of the historical context of the Act (*Epperson v. Arkansas*), the specific sequence of events leading up to passage of the Act, departures from normal procedural sequences, substantive departures from the normal (*Village of Arlington Heights v. Metropolitan Housing Corp.*), and contemporaneous statements of the legislative

14. Specifically, Senator Holsted testified that he holds to a literal interpretation of the Bible; that the bill was compatible with his religious beliefs; that the bill does favor the position of literalists; that his religious convictions were a factor in his sponsorship of the bill; and that he stated publicly to the *Arkansas Gazette* (although not on the floor of the Senate) contemporaneously with the legislative debate that the bill does presuppose the existence of a divine creator. There is no doubt that Senator Holsted knew he was sponsoring the teaching of a religious doctrine. His view was that the bill did not violate the First Amendment because, as he saw it, it did not favor one denomination over another.

15. This statute is, of course, clearly unconstitutional under the Supreme Court's decision in *Abbington School District v. Schempp*.

sponsor (*Federal Energy Administration v. Algonquin SNG, Inc.* at 564).

The unusual circumstances surrounding the passage of Act 590, as well as the substantive law of the First Amendment, warrant an inquiry into the stated legislative purposes. The author of the Act had publicly proclaimed the sectarian purpose of the proposal. The Arkansas residents who sought legislative sponsorship of the bill did so for a purely sectarian purpose. These circumstances alone may not be particularly persuasive, but when considered with the publicly announced motives of the legislative sponsor made contemporaneously with the legislative process; the lack of any legislative investigation, debate or consultation with any educators or scientists; the unprecedented intrusion in school curriculum;[16] and official history of the State of Arkansas on the subject; it is obvious that the statement of purposes has little, if any, support in fact. The State failed to produce any evidence which would warrant an inference or conclusion that at any point in the process anyone considered the legitimate educational value of the Act. It was simply and purely an effort to introduce the Biblical version of creation into the public school curricula. The only inference which can be drawn from these circumstances is that the Act was passed with the specific purpose by the General Assembly of advancing religion. The Act therefore fails the first prong of the three-pronged test, that of secular legislative purpose, as articulated in *Lemon v. Kurtzman* and *Stone v. Graham.*

III

If the defendants are correct and the Court is limited to an examination of the language of the Act, the evidence is overwhelming that both the purpose and effect of Act 590 is the advancement of religion in the public schools.

Section 4 of the Act provides:

Definitions. As used in this Act:

16. The joint stipulation of facts establishes that the following areas are the only *information* specifically required by statute to be taught in all Arkansas schools: (1) The effects of alcohol and narcotics on the human body, (2) Conservation of natural resources, (3) Bird Week, (4) Fire Prevention, and (5) Flag etiquette. Additionally, certain specific courses, such as American history and Arkansas history must be completed by each student before graduation from high school.

(a) "Creation-science" means the scientific evidences for creation and inferences from those scientific evidences. Creation-science includes the scientific evidences and related inferences that indicate: (1) Sudden creation of the universe, energy, and life from nothing; (2) The insufficiency of mutation and natural selection in bringing about development of all living kinds from a single organism; (3) Changes only within fixed limits of originally created kinds of plants and animals; (4) Separate ancestry for man and apes; (5) Explanation of the earth's geology by catastrophism, including the occurrence of a worldwide flood; and (6) A relatively recent inception of the earth and living kinds.

(b) "Evolution-science" means the scientific evidences for evolution and inferences from those scientific evidences. Evolution-science includes the scientific evidences and related inferences that indicate: (1) Emergence by naturalistic processes of the universe from disordered matter and emergence of life from nonlife; (2) The sufficiency of mutation and natural selection in bringing about development of present living kinds from simple earlier kinds; (3) Emergence by mutation and natural selection of present living kinds from simple earlier kinds; (4) Emergence of man from a common ancestor with apes; (5) Explanation of the earth's geology and the evolutionary sequence by uniformitarianism; and (6) An inception several billion years ago of the earth and somewhat later of life.

(c) "Public schools" means public secondary and elementary schools.

The evidence establishes that the definition of "creation science" contained in 4(a) has as its unmentioned reference the first 11 chapters of the Book of Genesis. Among the many creation epics in human history, the account of sudden creation from nothing, or *creatio ex nihilo,* and subsequent destruction of the world by flood is unique to Genesis. The concepts of 4(a) are the literal Fundamentalists' view of Genesis. Section 4(a) is unquestionably a statement of religion, with the exception of 4(a)(2) which is a negative thrust aimed at what the creationists understand to be the theory of evolution.[17]

Both the concepts and wording of Section 4(a) convey an inescapable religiosity. Section 4(a)(1) describes "sudden creation of the universe, energy and life from nothing." Every theologian who testified, including defense witnesses, expressed the opinion that the statement referred to a supernatural creation which was performed by God.

17. Paul Ellwanger stated in his deposition that he did not know why Section 4(a)(2) (insufficiency of mutation and natural selection) was included as an evidence supporting creation science. He indicated that he was not a scientist, "but these are the postulates that have been laid down by creation scientists" (Ellwanger Deposition, p. 136).

Defendants argue that: (1) the fact that 4(a) conveys ideas similar to the literal interpretation of Genesis does not make it conclusively a statement of religion; (2) that reference to a creation from nothing is not necessarily a religious concept since the Act only suggests a creator who has power, intelligence and a sense of design and not necessarily the attributes of love, compassion, and justice;[18] and (3) that simply teaching about the concept of a creator is not a religious exercise unless the student is required to make a commitment to the concept of a creator.

The evidence fully answers these arguments. The ideas of 4(a)(1) are not merely similar to the literal interpretation of Genesis; they are identical and parallel to no other story of creation.[19]

The argument that creation from nothing in 4(a)(1) does not involve a supernatural deity has no evidentiary or rational support. To the contrary, "creation out of nothing" is a concept unique to Western religions. In traditional Western religious thought, the conception of a creator of the world is a conception of God. Indeed, creation of the world "out of nothing" is the ultimate religious statement because God is the only actor. As Dr. Langdon Gilkey noted, the Act refers to one who has the power to bring all the universe into existence from nothing. The only "one" who has this power is God.[20]

18. Although defendants must make some effort to cast the concept of creation in non-religious terms, this effort surely causes discomfort to some of the Act's more theologically sophisticated supporters. The concept of a creator God distinct from the God of love and mercy is closely similar to the Marcion and Gnostic heresies, among the deadliest to threaten the early Christian church. These heresies had much to do with development and adoption of the Apostle's Creed as the official creedal statement of the Roman Catholic Church in the West. (Testimony of Gilkey.)

19. The parallels between Section 4(a) and Genesis are quite specific: (1) "sudden creation from nothing" is taken from Genesis 1:1–10 (Testimony of Vawter and Gilkey); (2) destruction of the world by flood of divine origin is a notion peculiar to Judeo-Christian tradition and is based on Chapters 7 and 8 of Genesis (Vawter); (3) the term "kinds" has no fixed scientific meaning, but appears repeatedly in Genesis (all scientific witnesses); (4) "relatively recent inception" means an age of the earth from 6,000 to 10,000 years and is based on the genealogy of the Old Testament using the rather astronomical ages assigned to the patriarchs (Gilkey and several of defendants' scientific witnesses); (5) Separate ancestry of man and ape focuses on the portion of the theory of evolution which Fundamentalists find most offensive (*Epperson v. Arkansas*).

20. "[C]oncepts concerning . . . a supreme being of some sort are manifestly religious . . . These concepts do not shed that religiosity merely because they are presented as philosophy or as science . . ."(*Malnak v. Yogi* at 1322).

The leading creationist writers, Morris and Gish, acknowledge that the idea of creation described in 4(a)(1) is the concept of creation by God and make no pretense to the contrary.[21] The idea of sudden creation from nothing, or *creatio ex nihilo*, is an inherently religious concept (Testimony of Vawter, Gilkey, Geisler, Ayala, Blount, and Hicks).

The argument advanced by defendants' witness Dr. Norman Geisler, that teaching the existence of God is not religious unless the teaching seeks a commitment, is contrary to common understanding and contradicts settled case law (*Stone v. Graham, Abbington School District v. Schempp*).

The facts that creation science is inspired by the Book of Genesis and that Section 4(a) is consistent with a literal interpretation of Genesis leave no doubt that a major effect of the Act is the advancement of particular religious beliefs. The legal impact of this conclusion will be discussed further at the conclusion of the Court's evaluation of the scientific merit of creation science.

IV(A)

The approach to teaching "creation science" and "evolution science" found in Act 590 is identical to the two-model approach espoused by the Institute for Creation Research and is taken almost verbatim from ICR writings. It is an extension of Fundamentalists' view that one must either accept the literal interpretation of Genesis or else believe in the godless system of evolution.

The two-model approach of the creationists is simply a contrived dualism[22] which has no scientific factual basis or legitimate educational

21. See, e.g., Plaintiffs' Exhibit 76, Morris *et al.*, *Scientific Creationism*, p. 203 (1980) ("If creation really is a fact, this means there is a *Creator*, and the universe is His creation"). Numerous other examples of such admissions can be found in the many exhibits which represent creationist literature, but no useful purpose would be served here by a potentially endless listing.

22. Morris, the Director of ICR and one who first advocated the two-model approach, insists that a true Christian cannot compromise with the theory of evolution and that the Genesis version of creation and the theory of evolution are mutually exclusive (Plaintiffs' Exhibit 31, Morris, *Studies in the Bible & Science*, pp. 102–103). The two-model approach was the subject of Dr. Richard Bliss's doctoral dissertation (Defense Exhibit 35). It is presented in Bliss, *Origins: Two Models—Evolution, Creation* (1978). Moreover, the two-model approach merely casts in educationalist language the dualism which appears in all creationist literature—creation (i.e., God) and evolution are presented as two alternative and mutually exclusive theories [See, e.g., Plaintiffs' Exhibit 75, Morris,

purpose. It assumes only two explanations for the origins of life and existence of man, plants, and animals: It was either the work of a creator or it was not. Application of these two models, according to creationists, and the defendants, dictates that all scientific evidence which fails to support the theory of evolution is necessarily scientific evidence in support of creationism and is, therefore, creation science "evidence" in support of Section 4(a).

IV(B)

The emphasis on origins as an aspect of the theory of evolution is peculiar to creationist literature. Although the subject of origins of life is within the province of biology, the scientific community does not consider origins of life a part of evolutionary theory. The theory of evolution assumes the existence of life and is directed to an explanation of *how* life evolved. Evolution does not presuppose the absence of a creator or God and the plain inference conveyed by Section 4 is erroneous.[23]

As a statement of the theory of evolution, Section 4(b) is simply a hodgepodge of limited assertions, many of which are factually inaccurate.

For example, although 4(b)(2) asserts, as a tenet of evolutionary theory, "the sufficiency of mutation and natural selection in bringing about the existence of present living kinds from simple earlier kinds," Drs. Ayala and Gould both stated that biologists know that these two processes do not account for all significant evolutionary change. They testified to such phenomena as recombination, the founder effect, genetic drift and the theory of punctuated equilibrium, which are believed to play important evolutionary roles. Section 4(b) omits any reference to these. Moreover, 4(b) utilizes the term "kinds" which all scientists said

---

Scientific Creationism (1974) (public school edition); Plaintiffs' Exhibit 59, Fox, *Fossils: Hard Facts from the Earth*]. Particularly illustrative is Plaintiffs' Exhibit 61, Boardman, et al., *Worlds Without End* (1971), a CSRC publication: "One group of scientists, known as creationists, believe that God, in a miraculous manner, created all matter and energy. . . . "Scientists who insist that the universe just grew, by accident, from a mass of hot gases without the direction or help of a Creator are known as evolutionists."

23. The idea that belief in a creator and acceptance of the scientific theory of evolution are mutually exclusive is a false premise and offensive to the religious views of many (Testimony of Hicks). Dr. Francisco Ayala, a geneticist of considerable renown and a former Catholic priest who has the equivalent of a Ph.D. in theology, pointed out that many working scientists who subscribe to the theory of evolution are devoutly religious.

is not a word of science and has no fixed meaning. Additionally, the Act presents both evolution and creation science as "package deals." Thus, evidence critical of some aspect of what the creationists define as evolution is taken as support for a theory which includes a worldwide flood and a relatively young earth.[24]

IV(C)

In addition to the fallacious pedagogy of the two-model approach, Section 4(a) lacks legitimate educational value because "creation science" as defined in that section is simply not science. Several witnesses suggested definitions of science. A descriptive definition was said to be that science is what is "accepted by the scientific community" and is "what scientists do." The obvious implication of this description is that, in a free society, knowledge does not require the imprimatur of legislation in order to become science.

More precisely, the essential characteristics of science are:

1. It is guided by natural law;
2. It has to be explanatory by reference to natural law;
3. It is testable against the empirical world;
4. Its conclusions are tentative, i.e., are not necessarily the final word; and
5. It is falsifiable (Testimony of Ruse and other science witnesses).

Creation science as described in Section 4(a) fails to meet these essential characteristics. First, the section revolves around 4(a)(1) which asserts a sudden creation "from nothing." Such a concept is not science because it depends upon a supernatural intervention which is not guided by natural law. It is not explanatory by reference to natural law, is not testable, and is not falsifiable.[25]

---

24. This is so despite the fact that some of the defense witnesses do not subscribe to the young earth or flood hypotheses. Dr. Geisler stated his belief that the earth is several billion years old. Dr. Wickramasinghe stated that no rational scientist would believe the earth is less than one million years old or that all the world's geology could be explained by a worldwide flood.

25. "We do not know how God created, what processes He used, for *God used processes which are not now operating anywhere in the natural universe.* This is why we refer to divine creation as Special Creation. We cannot discover by scientific investigation anything about the creative processes used by God" [Plaintiffs' Exhibit 78, Gish, *Evolution? The Fossils Say No!*, p. 42 (3rd ed., 1979) (emphasis in original)].

If the unifying idea of supernatural creation by God is removed from Section 4, the remaining parts of the section explain nothing and are meaningless assertions.

Section 4(a)(2), relating to the "insufficiency of mutation and natural selection in bringing about development of all living kinds from a single organism," is an incomplete negative generalization directed at the theory of evolution.

Section 4(a)(3) which describes "changes only within fixed limits of originally created kinds of plants and animals" fails to conform to the essential characteristics of science for several reasons. First, there is no scientific definition of "kinds" and none of the witnesses was able to point to any scientific authority which recognized the term or knew how many "kinds" existed. One defense witness suggested there may be 100 to 10,000 different "kinds." Another believes there were "about 10,000, give or take a few thousand." Second, the assertion appears to be an effort to establish outer limits of changes within species. There is no scientific explanation for these limits which is guided by natural law and the limitations, whatever they are, cannot be explained by natural law.

The statement in 4(a)(4) of "separate ancestry of man and apes" is a bald assertion. It explains nothing and refers to no scientific fact or theory.[26]

Section 4(a)(5) refers to "explanation of the earth's geology by catastrophism, including the occurrence of a worldwide flood." This assertion completely fails as science. The Act is referring to the Noachian flood described in the Book of Genesis.[27] The creationist writers concede that *any* kind of Genesis Flood depends upon supernatural intervention. A worldwide flood as an explanation of the world's geology is not the

---

26. The evolutionary notion that man and some modern apes have a common ancestor somewhere in the distant past has consistently been distorted by anti-evolutionists to say that man descended from modern monkeys. As such, this idea has long been most offensive to Fundamentalists (See *Epperson v. Arkansas*).

27. Not only was this point acknowledged by virtually all the defense witnesses, it is patent in the creationist literature. See, e.g., Plaintiffs' Exhibit 89, Kofahl & Segraves, *The Creation Explanation*, p. 40: "The Flood of Noah brought about vast changes in the earth's surface, including vulcanism, mountain building, and the deposition of the major part of sedimentary strata. This principle is called 'Biblical catastrophism.' "

product of natural law, nor can its occurrence be explained by natural law.

Section 4(a)(6) equally fails to meet the standards of science. "Relatively recent inception" has no scientific meaning. It can only be given meaning by reference to creationist writings which place the age at between 6,000 and 20,000 years because of the genealogy of the Old Testament. See, e.g., Plaintiffs' Exhibit 78, Gish (6,000 to 10,000); Plaintiffs' Exhibit 87, Segraves (6,000 to 20,000). Such a reasoning process is not the product of natural law; not explainable by natural law; nor is it tentative.

Creation science, as defined in Section 4(a), not only fails to follow the canons defining scientific theory, it also fails to fit the more general descriptions of "what scientists think" and "what scientists do." The scientific community consists of individuals and groups, nationally and internationally, who work independently in such varied fields as biology, paleontology, geology and astronomy. Their work is published and subject to review and testing by their peers. The journals for publication are both numerous and varied. There is, however, not one recognized scientific journal which has published an article espousing the creation science theory described in Section 4(a). Some of the State's witnesses suggested that the scientific community was "close-minded" on the subject of creationism and that explained the lack of acceptance of the creation science arguments. Yet no witness produced a scientific article for which publication had been refused. Perhaps some members of the scientific community are resistant to new ideas. It is, however, inconceivable that such a loose knit group of independent thinkers in all the varied fields of science could, or would, so effectively censor new scientific thought.

The creationists have difficulty maintaining among their ranks consistency in the claim that creationism is science. The author of Act 590, Ellwanger, said that neither evolution nor creationism was science. He thinks both are religion. Duane Gish recently responded to an article in *Discover* critical of creationism by stating:

Stephen Jay Gould states that creationists claim creation is a scientific theory. This is a false accusation. Creationists have repeatedly stated that neither creation nor evolution is a scientific theory (and each is equally religious). (Gish, letter to the editor of *Discover*, July 1981, Appendix 30 to Plaintiffs' Pretrial Brief.)

The methodology employed by creationists is another factor which is indicative that their work is not science. A scientific theory must be tentative and always subject to revision or abandonment in light of facts that are inconsistent with, or falsify, the theory. A theory that is by its own terms dogmatic, absolutist, and never subject to revision is not a scientific theory.

The creationists' methods do not take data, weigh it against the opposing scientific data, and thereafter reach the conclusions stated in Section 4(a). Instead, they take the literal wording of the Book of Genesis and attempt to find scientific support for it. The method is best explained in the language of Morris in his book *Studies in the Bible and Science* at page 114 (Plaintiffs' Exhibit 31):

... it is .. quite impossible to determine anything about Creation through a study of present processes, because present processes are not creative in character. If man wishes to know anything about Creation (the time of Creation, the duration of Creation, the order of Creation, the methods of Creation, or anything else) his sole source of true information is that of divine revelation. God was there when it happened. We were not there . . . Therefore, we are completely limited to what God has seen fit to tell us, and this information is in His written Word. This is our textbook on the science of Creation! (Morris, Plaintiffs' Exhibit 31.)

The Creation Research Society employs the same unscientific approach to the issue of creationism. Its applicants for membership must subscribe to the belief that the Book of Genesis is "historically and scientifically true in all of the original autographs."[28] The Court would never criticize or discredit any person's testimony based on his or her religious beliefs. While anybody is free to approach a scientific inquiry in any fashion they choose, they cannot properly describe the methodology used as scientific, if they start with a conclusion and refuse to change it regardless of the evidence developed during the course of the investigation.

IV(D)

In efforts to establish "evidence" in support of creation science, the defendants relied upon the same false premise as the two-model approach contained in Section 4, i.e., all evidence which criticized evolutionary theory was proof in support of creation science. For example, the defendants established that the mathematical probability of a chance chemical combination resulting in life from non-life is so remote that such an occurrence is almost beyond imagination. Those mathematical

---

28. See note 7, *supra*, for the full text of the CRS creed.

facts, the defendants argue, are scientific evidences that life was the product of a creator. While the statistical figures may be impressive evidence against the theory of chance chemical combinations as an explanation of origins, it requires a leap of faith to interpret those figures so as to support a complex doctrine which includes a sudden creation from nothing, a worldwide flood, separate ancestry of man and apes, and a young earth.

The defendants' argument would be more persuasive if, in fact, there were only two theories or ideas about the origins of life and the world. That there are a number of theories was acknowledged by the State's witnesses, Dr. Wickramasinghe and Dr. Geisler. Dr. Wickramasinghe testified at length in support of a theory that life on earth was "seeded" by comets which delivered genetic material and perhaps organisms to the earth's surface from interstellar dust far outside the solar system. The "seeding" theory further hypothesizes that the earth remains under the continuing influence of genetic material from space which continues to affect life. While Wickramasinghe's theory[29] about the origins of life on earth has not received general acceptance within the scientific community, he has, at least, used scientific methodology to produce a theory of origins which meets the essential characteristics of science.

Perhaps Dr. Wickramasinghe was called as a witness because he was generally critical of the theory of evolution and the scientific community, a tactic consistent with the strategy of the defense. Unfortunately for the defense, he demonstrated that the simplistic approach of the two-model analysis of the origins of life is false. Furthermore, he corroborated the plaintiffs' witnesses by concluding that "no rational scientist" would believe the earth's geology could be explained by reference to a world-wide flood or that the earth was less than one million years old.

The proof in support of creation science consisted almost entirely of efforts to discredit the theory of evolution through a rehash of data and theories which have been before the scientific community for decades. The arguments asserted by creationists are not based upon new scientific evidence or laboratory data which has been ignored by the scientific community.

Robert Gentry's discovery of radioactive polonium haloes in granite and coalified woods is, perhaps, the most recent scientific work which the creationists use as argument for a "relatively recent inception" of

29. The theory is detailed in Wickramasinghe's book with Sir Fred Hoyle, *Evolution From Space* (1981), which is Defense Exhibit 79.

the earth and a "worldwide flood." The existence of polonium haloes in granite and coalified wood is thought to be inconsistent with radiometric dating methods based upon constant radioactive decay rates. Mr. Gentry's findings were published almost ten years ago and have been the subject of some discussion in the scientific community. The discoveries have not, however, led to the formulation of any scientific hypothesis or theory which would explain a relatively recent inception of the earth or a worldwide flood. Gentry's discovery has been treated as a minor mystery which will eventually be explained. It may deserve further investigation, but the National Science Foundation has not deemed it to be of sufficient import to support further funding.

The testimony of Marianne Wilson was persuasive evidence that creation science is not science. Ms. Wilson is in charge of the science curriculum for Pulaski County Special School District, the largest school district in the State of Arkansas. Prior to the passage of Act 590, Larry Fisher, a science teacher in the District, using materials from the ICR, convinced the School Board that it should voluntarily adopt creation science as part of its science curriculum. The District Superintendent assigned Ms. Wilson the job of producing a creation science curriculum guide. Ms. Wilson's testimony about the project was particularly convincing because she obviously approached the assignment with an open mind and no preconceived notions about the subject. She had not heard of creation science until about a year ago and did not know its meaning before she began her research.

Ms. Wilson worked with a committee of science teachers appointed from the District. They reviewed practically all of the creationist literature. Ms. Wilson and the committee members reached the unanimous conclusion that creationism is not science; it is religion. They so reported to the Board. The Board ignored the recommendation and insisted that a curriculum guide be prepared.

In researching the subject, Ms. Wilson sought the assistance of Mr. Fisher who initiated the Board action and asked professors in the science departments of the University of Arkansas at Little Rock and the University of Central Arkansas[30] for reference material and assistance, and attended a workshop conducted at Central Baptist College by Dr. Rich-

30. Ms. Wilson stated that some professors she spoke with sympathized with her plight and tried to help her find scientific materials to support Section 4(a). Others simply asked her to leave.

ard Bliss of the ICR staff. Act 590 became law during the course of her work so she used Section 4(a) as a format for her curriculum guide.

Ms. Wilson found all available creationists' materials unacceptable because they were permeated with religious references and reliance upon religious beliefs.

It is easy to understand why Ms. Wilson and other educators find the creationists' textbook material and teaching guides unacceptable. The materials misstate the theory of evolution in the same fashion as Section 4(b) of the Act, with emphasis on the alternative mutually exclusive nature of creationism and evolution. Students are constantly encouraged to compare and make a choice between the two models, and the material is not presented in an accurate manner.

A typical example is *Origins* (Plaintiffs' Exhibit 76) by Richard B. Bliss, Director of Curriculum Development of the ICR. The presentation begins with a chart describing "preconceived ideas about origins" which suggests that some people believe that evolution is atheistic. Concepts of evolution, such as "adaptive radiation," are erroneously presented. At page 11, Figure 1.6, of the text, a chart purports to illustrate this "very important" part of the evolution model. The chart conveys the idea that such diverse mammals as a whale, bear, bat and monkey all evolved from a shrew through the process of adaptive radiation. Such a suggestion is, of course, a totally erroneous and misleading application of the theory. Even more objectionable, especially when viewed in light of the emphasis on asking the student to elect one of the models, is the chart presentation at page 17, Figure 1.6. That chart purports to illustrate the evolutionists' belief that man evolved from bacteria to fish to reptile to mammals and, thereafter, into man. The illustration indicates, however, that the mammal from which man evolved was *a rat.*

Biology, A Search For Order in Complexity[31] is a high school biology text typical of creationists' materials. The following quotations are illustrative:

Flowers and roots do not have a mind to have purpose of their own; therefore, this planning must have been done for them by the Creator. (p. 12)

31. Plaintiffs' Exhibit 129, published by Zonderman Publishing House (1974), states that it was "prepared by the Textbook Committee of the Creation Research Society." It has a disclaimer pasted inside the front cover stating that it is not suitable for use in public schools.

The exquisite beauty of color and shape in flowers exceeds the skill of poet, artist, and king. Jesus said (from Matthew's gospel), 'Consider the lilies of the field, how they grow; they toil not, neither do they spin . . . .(p. 363)

The "public school edition" texts written by creationists simply omit Biblical references but the content and message remain the same. For example, *Evolution? The Fossils Say No!*[32] contains the following:

Creation. By creation we mean the bringing into being by a supernatural Creator of the basic kinds of plants and animals by the process of sudden, or fiat, creation.
We do not know how the Creator created, what processes He used, *for He used processes which are not now operating anywhere in the natural universe.* This is why we refer to creation as Special Creation. We cannot discover by scientific investigation anything about the creative processes used by the Creator. (p. 40)

Gish's book also portrays the large majority of evolutionists as "materialistic atheists or agnostics."

Scientific Creationism (Public School Edition) by Morris, is another text reviewed by Ms. Wilson's committee and rejected as unacceptable. The following quotes illustrate the purpose and theme of the text:

Foreword. Parents and youth leaders today, and even many scientists and educators, have become concerned about the prevalence and influence of evolutionary philosophy in modern curriculum. Not only is this system inimical to orthodox Christianity and Judaism, but also, as many are convinced, to a healthy society and true science as well. (p. iii)

The rationalist of course finds the concept of special creation insufferably naive, even "incredible." Such a judgment, however, is warranted only if one categorically dismisses the existence of an omnipotent God. (p. 17)

Without using creationist literature, Ms. Wilson was unable to locate one genuinely scientific article or work which supported Section 4(a). In order to comply with the mandate of the Board she used such materials as an article from *Reader's Digest* about "atomic clocks" which inferentially suggested that the earth was less than 4-1/2 billion years old. She was unable to locate any substantive teaching material for some parts of Section 4 such as the worldwide flood. The curriculum guide which she prepared cannot be taught and has no educational value as science. The defendants did not produce any text or writing

32. Plaintiffs' Exhibit 77, by Duane Gish.

in response to this evidence which they claimed was usable in the public school classroom.[33]

The conclusion that creation science has no scientific merit or educational value as science has legal significance in light of the Court's previous conclusion that creation science has, as one major effect, the advancement of religion. The second part of the three-pronged test for establishment reaches only those statutes having as their *primary* effect the advancement of religion. Secondary effects which advance religion are not constitutionally fatal. Since creation science is not science, the conclusion is inescapable that the *only* real effect of Act 590 is the advancement of religion. The Act therefore fails both the first and second portions of the test in *Lemon v. Kurtzman.*

IV(E)

Act 590 mandates "balanced treatment" for creation science and evolution science. The Act prohibits instruction in any religious doctrine or references to religious writings. The Act is self-contradictory and compliance is impossible unless the public schools elect to forego significant portions of subjects such as biology, world history, geology, zoology, botany, psychology, anthropology, sociology, philosophy, physics and chemistry. Presently, the concepts of evolutionary theory as described in 4(b) permeate the public school textbooks. There is no way teachers can teach the Genesis account of creation in a secular manner.

The State Department of Education, through its textbook selection committee, school boards, and school administrators will be required to constantly monitor materials to avoid using religious references. The school boards, administrators and teachers face an impossible task. How is the teacher to respond to questions about a creation suddenly and out of nothing? How will a teacher explain the occurrence of a

---

33. The passage of Act 590 apparently caught a number of its supporters off guard as much as it did the school district. The Act's author, Paul Ellwanger, stated in a letter to "Dick" (apparently Dr. Richard Bliss at ICR): "And finally, if you know of any textbooks at any level and for any subjects that you think are acceptable to you and also constitutionally admissible, these are things that would be of *enormous* to these bewildered folks who may be caught, as Arkansas now has been, by the sudden need to implement a whole new ball game with which they are quite unfamiliar." (sic) (Unnumbered attachment to Ellwanger Deposition.)

worldwide flood? How will a teacher explain the concept of a relatively recent age of the earth? The answer is obvious because the only source of this information is ultimately contained in the Book of Genesis.

References to the pervasive nature of religious concepts in creation science texts amply demonstrate why State entanglement with religion is inevitable under Act 590. Involvement of the State in screening texts for impermissible religious references will require State officials to make delicate religious judgments. The need to monitor classroom discussion in order to uphold the Act's prohibition against religious instruction will necessarily involve administrators in questions concerning religion. These continuing involvements of State officials in questions and issues of religion create an excessive and prohibited entanglement with religion (*Brandon v. Board of Education* at 1230).

V

These conclusions are dispositive of the case and there is no need to reach legal conclusions with respect to the remaining issues. The plaintiffs raised two other issues questioning the constitutionality of the Act and, insofar as the factual findings relevant to these issues are not covered in the preceding discussion, the Court will address these issues. Additionally, the defendants raised two other issues which warrant discussion.

V(A)

First, plaintiff teachers argue the Act is unconstitutionally vague to the extent that they cannot comply with its mandate of "balanced" treatment without jeopardizing their employment. The argument centers around the lack of a precise definition in the Act for the word "balanced." Several witnesses expressed opinions that the word has such meanings as equal time, equal weight, or equal legitimacy. Although the Act could have been more explicit, "balanced" is a word subject to ordinary understanding. The proof is not convincing that a teacher using a reasonably acceptable understanding of the word and making a good faith effort to comply with the Act will be in jeopardy of termination. Other portions of the Act are arguably vague, such as the "relatively recent" inception of the earth and life. The evidence establishes, however, that relatively recent means from 6,000 to 20,000 years, as commonly understood in creation science literature. The meaning of this phrase, like Section 4(a) generally, is, for purposes of the Establishment Clause, all too clear.

V(B)

The plaintiffs' other argument revolves around the alleged infringement by the defendants upon the academic freedom of teachers and students. It is contended this unprecedented intrusion in the curriculum by the State prohibits teachers from teaching what they believe should be taught or requires them to teach that which they do not believe is proper. The evidence reflects that traditionally the State Department of Education, local school boards, and administration officials exercise little, if any, influence upon the subject matter taught by classroom teachers. Teachers have been given freedom to teach and emphasize those portions of subjects the individual teacher considered important. The limits to this discretion have generally been derived from the approval of textbooks by the State Department and preparation of curriculum guides by the school districts.

Several witnesses testified that academic freedom for the teacher means, in substance, that the individual teacher should be permitted unlimited discretion subject only to the bounds of professional ethics. The Court is not prepared to adopt such a broad view of academic freedom in the public schools.

In any event, if Act 590 is implemented, many teachers will be required to teach material in support of creation science which they do not consider academically sound. Many teachers will simply forego teaching subjects which might trigger the "balanced treatment" aspects of Act 590 even though they think the subjects are important to a proper presentation of a course.

Implementation of Act 590 will have serious and untoward consequences for students, particularly those planning to attend college. Evolution is the cornerstone of modern biology, and many courses in public schools contain subject matter relating to such varied topics as the age of the earth, geology, and relationships among living things. Any student who is deprived of instruction as to the prevailing scientific thought on these topics will be denied a significant part of science education. Such a deprivation through the high school level would undoubtedly have an impact upon the quality of education in the State's colleges and universities, especially including the pre-professional and professional programs in the health sciences.

V(C)

The defendants argue in their brief that evolution is, in effect, a religion, and that by teaching a religion which is contrary to some students' religious view, the State is infringing upon the student's free exercise rights under the First Amendment. Mr. Ellwanger's legislative findings, which were adopted as a finding of fact by the Arkansas Legislature in Act 590, provides:

Evolution-science is contrary to the religious convictions or moral values or philosophical beliefs of many students and parents, including individuals of many different religious faiths and with diverse moral and philosophical beliefs [Act 590, §7(d)].

The defendants argue that the teaching of evolution alone presents both a free exercise problem and an establishment problem which can only be redressed by giving balanced treatment to creation science, which is admittedly consistent with some religious beliefs. This argument appears to have its genesis in a student note written by Mr. Wendell Bird, "Freedom of Religion and Science Instruction in Public Schools," 87 *Yale Law Journal* 515 (1978). The argument has no legal merit.

If creation science is, in fact, science and not religion, as the defendants claim, it is difficult to see how the teaching of such a science could "neutralize" the religious nature of evolution.

Assuming for the purposes of argument, however, that evolution is a religion or religious tenet, the remedy is to stop the teaching of evolution; not establish another religion in opposition to it. Yet it is clearly established in the case law, and perhaps also in common sense, that evolution is not a religion and that teaching evolution does not violate the Establishment Clause (*Epperson v. Arkansas, Willoughby v. Stever, Wright v. Houston Independent School District*).

V(D)

The defendants presented Dr. Larry Parker, a specialist in devising curricula for public schools. He testified that the public school's curriculum should reflect the subjects the public wants taught in schools. The witness said that polls indicated a significant majority of the American public thought creation science should be taught if evolution was taught. The point of this testimony was never placed in a legal context. No doubt a sizeable majority of Americans believe in the concept of a Creator or, at least, are not opposed to the concept and see nothing wrong with teaching school children about the idea.

The application and content of First Amendment principles are not determined by public opinion polls or by a majority vote. Whether the proponents of Act 590 constitute the majority or the minority is quite irrelevant under a constitutional system of government. No group, no matter how large or small, may use the organs of government, of which the public schools are the most conspicuous and influential, to foist its religious beliefs on others.

The Court closes this opinion with a thought expressed eloquently by the great Justice Frankfurter:

We renew our conviction that 'we have staked the very existence of our country on the faith that complete separation between the state and religion is best for the state and best for religion' (*Everson v. Board of Education* at 59). If nowhere else, in the relation between Church and State, 'good fences make good neighbors' (*McCollum v. Board of Education* at 232).

An injunction will be entered permanently prohibiting enforcement of Act 590.

It is so ordered this January 5, 1982.

William R. Overton
United States District Judge

Cases Cited

Abbington School District v. Schempp, 374 U.S. 203 (1963).

Brandon v. Board of Education, 487 F. Supp. 1219 (N.D.N.Y.), *affirmed*, 635 F.2d 971 (2nd Circuit, 1980).

Chrysler Corporation v. Brown, 441 U.S. 281 (1979).

Committee for Public Education & Religious Liberty v. Nyquist, 413 U.S. 756 (1973).

Engel v. Vitale, 370 U.S. 421 (1962).

Epperson v. Arkansas, 393 U.S. 97 (1968).

Everson v. Board of Education, 330 U.S. 1 (1947).

Federal Energy Administration v. Algonquin SNG, Inc., 426 U.S. 548 (1976).

Hans v. Louisiana, 134 U.S. 1 (1890).

Lemon v. Kurtzman, 403 U.S. 602 (1971).

McCollum v. Board of Education, 333 U.S. 203 (1948).

McGowan v. Maryland, 366 U.S. 420 (1961).

Malnak v. Yogi, 440 F. Supp. 1284 (D.N.J. 1977), *affirmed per curiam,* 592 F.2d 197 (3rd Circuit 1979).

Stone v. Graham, 449 U.S. 39 (1980).

United States v. Emmons, 410 U.S. 396 (1979).

Village of Arlington Heights v. Metropolitan Housing Corp., 429 U.S. 252 (1977).

Willoughby v. Stever, No. 15574–75 (Denver District Court, 18 May 1973), *affirmed* 504 F.2d 271 (D.C. Circuit 1974), *certiorari denied,* 420 U.S. 927 (1975).

Wright v. Houston Independent School District, 366 F. Supp. 1208 (Southern District of Texas, 1978), *affirmed,* 486 F.2d 137 (5th Circuit 1973), *certiorari denied,* 417 U.S. 969 (1974).

From Dayton to Little Rock: Creationism Evolves*   Dorothy Nelkin

The 1981 legislation in Arkansas and Louisiana which required "balanced treatment of creation-science and evolution-science" represents the most ambitious effort of the "scientific creationists" to date to gain equal time for the teaching of the book of Genesis as an alternative and viable scientific theory of origins. The trial testing the constitutionality of the Arkansas law culminated in a powerful and unambiguous decision; however, creationists continue to lobby for similar legislation in many other states. Far from an aberration, today's scientific creationists represent a modern manifestation of a deep fundamentalist current in American social history.

Origins

The American social movement known as Fundamentalism developed during the late nineteenth century as a defensive response to the threatening social changes brought about by immigration and the Industrial Revolution. Literal interpretation of the Bible became a bulwark against modern ideas. The movement, rooted in Methodist Revivalism in the early 1800s, began to crystallize around the issue of Darwinism, which was perceived as a threat to Christian beliefs.[1] By the 1920s, the issue was dividing Protestant churches as fundamentalists denied the validity of evolution and modernists sought to reconcile their faith with science. Control over educational institutions became the focus of their battles.

At this time, the fundamentalists' reaction against evolution came mainly from "backwoods" areas, from rural people irritated by the

*This essay is drawn from Dorothy Nelkin, *The Creation Controversy: Science or Scriptures in the Schools* (New York: Norton, 1982).

liberal attitudes of the industrial North. Indeed, among most Northerners, any assumption about the incompatability of science and religion seemed absurd by the time the Scopes trial[2] came to national prominence. In fact, the trial of the Tennessee high school biology teacher, John T. Scopes, was not intended to raise that issue at all. Rather, it was provoked by the American Civil Liberties Union (ACLU) in order to show that Tennessee's 1925 anti-evolution legislation violated the First Amendment. The constitutional question, however, became buried as William Jennings Bryan and Clarence Darrow clashed instead over questions of science, religion, and morality.

Arthur Garfield Hays, who worked on the Scopes' defense team with Darrow, could hardly believe that "religious views of the middle ages" could recur "in spite of railroads, steam boats, the World War, the telephone, the radio, the airplane, all the great mechanistic discoveries. . . ."[3] Yet fundamentalists had considerable political influence in the 1920s. Between 1921 and 1929, they introduced anti-evolution bills into thirty-seven state legislatures. Bills were passed in three: Mississippi (1926), Arkansas (1928), and Texas (1929). By threatening to exclude such textbooks from adoption in schools, fundamentalists also persuaded publishers to qualify statements about evolution in textbooks that included discussion of the theory.

A contemporary writer described the prevailing emotional hostility to science as a "cancer of ignorance,"[4] a repudiation of the authority and integrity of science. That hostility was fueled by the association of evolution theory with disturbing contemporary social problems. Evolution ideology, claimed Bryan, went beyond simple scientific questions and bore on moral values—a forceful argument in the 1920s, when popular discussion suggested that the country was "going to ruin." Social problems were variously attributed to postwar letdown, to prosperity, and to the weakening of loyalty to the church; but blame for "immorality" was also laid on a materialism fed by science and, especially, the teaching of evolution.

The Scopes trial also reflected parental desire to control children's education and values. "The question is," Bryan asked, "can a minority in this State come in and compel a teacher to teach that the Bible is not true and make the parents of these children pay the expenses of the teacher to tell their children what these people believe is false and dangerous?"[5] Scientists, of course, had quite a different view: "What is to be taught as science would be determined not by a consensus of the best scientific opinion, but by the votes of shopgirls and farm-hands, ignorant alike of science and of the foundation principles of our civil society."[6]

Although John Scopes lost his case, the trial suspended the public efforts of fundamentalists to ban the teaching of evolution. During the Depression, as economic problems and Prohibition preoccupied fundamentalist leaders, they backed away from textbook crusades. During the 1940s and 1950s, fundamentalists worked to reinforce their own subculture, founding Bible camps, colleges, seminaries, and newspapers, and purchasing radio stations. To the extent that they attacked the public schools, they focused more on prayers and sex education than on evolution.

The relative quiescence of fundamentalists at this time also reflected the neglect of evolution in biology textbooks after the Scopes trial. Textbooks published throughout the late 1920s ignored evolution theory, and new editions of older volumes deleted the word *evolution* and the name *Darwin* from their indexes. As late as 1959, Hermann J. Muller wrote that "antiquated religious traditions" dominated the teaching of biology in public schools.[7] One hundred years after the theory of evolution by natural selection was firmly established, it was still not an integral part of the public school curriculum.

Meanwhile, legislation forbidding the teaching of evolution in public schools remained in force in several Southern states. In January 1961, a bill to repeal the Tennessee statute met prompt and passionate rejection by legislators, who argued that evolution theory "drives God out of the universe" and "leads to communism."[8] Not until 1967, when Gary Scott, a Tennessee high school teacher who had been dismissed from his job for violating the state statute, challenged the law, was it finally repealed.[9] A few years earlier, in 1965, another high school teacher, Susan Epperson, had challenged the constitutionality of the 1928 Arkansas law.[10] Although Epperson won a favorable judgment in a local court, the Arkansas Supreme Court upheld the law. In 1968, however, the U.S. Supreme Court ruled the Arkansas anti-evolution law unconstitutional, and following this precedent, the last of the 1920s laws (in Mississippi) was soon repealed.

Repeal of these anti-evolution laws did not mean that social values had changed. Even as the courts were challenging the old legislation, fundamentalist textbook watchers were gaining momentum for a renewed attack on the evolutionary assumptions of the new biology courses being introduced in public schools.

Creationism Evolved

Creationists today resemble their fundamentalist ancestors of the 1920s in many respects. They accept the Biblical doctrine of creation as literal: "all basic types of living things, including man, were made by a direct creative act of God during the creation week described in *Genesis*."[11] They believe that creation theory is the most basic of all Christian beliefs ("at the very center of the warfare . . . against Satan") and regard the failure by many churches to emphasize special creation as "a tragic oversight that has resulted in defection . . . to the evolutionary world view and then inevitably later to liberalism."[12] They associate evolution theory with immorality. Judge Braswell Deen, a creationist and judge of the Georgia State Court of Appeals, has written that the "monkey mythology of Darwin" causes "permissiveness, promiscuity, pills, prophylactics, perversions, pregnancies, abortions, pornotherapy, pollution, poisoning, and proliferation of crimes."[13] Like their 1920s' predecessors, these fundamentalists resent "the elite corps of unelected professional academics and their government friends [who] run things in the schools."[14]

The difference is that, at the same time they are reinterpreting the theory of origins in terms of Biblical authority, today's creationists present themselves as scientists, reinterpreting fossil evidence from the perspective of special creation. They are not backwoods rural folk, but solid, middle-class, often technically trained people working in high-technology professions in centers of science-based industry. These modern-day creationists share many of the moral and religious concerns expressed in the 1920s, but adopt a style strikingly different from that of their flamboyant ancestors.

In contrast to the ballyhoo in Dayton, Tennessee, creationist confrontations today are more like debates within professional societies. Creationists argue that the Book of Genesis is not religious dogma, but an alternative scientific hypothesis capable of evaluation by scientific procedures. They present themselves not as believers, but as scientists engaged in scholarly debate about the methodological validity of two scientific theories. And they seek to implement their views through school boards, textbook commissions, state legislatures, and the courts.

Tactics

In California, the official *Science Framework* provides the guidelines for the public school science program. It is implemented by a state curriculum commission that advises the Board of Education on the choice of textbooks to receive state subsidies. In November 1969, creationists in California persuaded the Board of Education to include a brief statement in the *Science Framework* that "creation and evolutionary theories are not necessarily mutually exclusives. Some of the scientific data . . . may be best explained by creation theory, while other data . . . substantiate a process of evolution. . . ."[15] The implication of these new guidelines became evident in 1971 when the Commission began to negotiate with publishers to adapt their texts. Faced with growing and unresolvable controversy, the Board of Education held a hearing in November 1972 to assess public opinion. As the first public confrontation between creationists and evolutionists since the Scopes trial, it promised to be a circus—but bureaucratic procedures and the creationists' efforts to present themselves as scientists set a tone of sober debate. In fact, the ironies were striking: the 23 witnesses for the creationist position included only a few ministers, and 12 scientists and engineers. The evolutionists called forth only four scientists and the rest were ministers, rabbis, and priests. However, in the creationists' testimony, the old themes prevailed: as a threat to traditional Christian values, the teaching of evolution was undermining the morality of American youth. Vernon Grose, the man who had written the creationist revision of the *Science Framework*, described evolution theory as "a campaign of secularization in a scientific-materialistic society—a campaign to totally neutralize religion's convictions, to destroy any concept of absolute moral values. . . ."[16]

Following this hearing, the California Board of Education recommended relatively innocuous changes in biology textbooks to remove "dogmatism" and to guard against "the religion of science" as well as "other" religious positions. As a further concession to creationists, the Board agreed to send a memo to school districts, reminding teachers to include conditional statements whenever discussing the origins of life.

Creationists felt sold out in California. Nevertheless, the publicity expanded their constituency and strengthened their resolve to influence school boards and textbook commissions in many other states. These efforts at the local and state level have proved to be the creationists' major source of influence, for publishers have been willing, if not eager, to adapt to public pressure. The existence of potential large markets (especially in the conservative sunbelt states) has caused publishers today, just as in the 1930s, to accommodate to creationist pressures by adding qualifications to statements about evolution theory or by simply avoiding sensitive issues. Some delete references to fossil formation, geological eras, and the age of the earth. Many textbooks include discussion of evolution only in a single chapter that the reader can avoid.

Legal battles for creationists not only give them wider publicity but also help to develop loyalty among their supporters. In some cases, creationists themselves have brought cases to court, asserting their First Amendment rights. In August 1972, William Willoughby, religion editor of the *Washington Evening Star*, filed suit ("in the interest of forty million evangelistic Christians in the United States") against the Director of the National Science Foundation (NSF) and the Board of Regents of the University of Colorado.[17] The NSF had supported a Colorado project to develop biology textbooks, and Willoughby wanted the Foundation to spend an equal amount "for the promulgation of the creationist theory of the evolution of man." Willoughby alleged that citizens are coerced to pay taxes to support educational activities that offend their religious beliefs. Supporting educational programs that are "one-sided, biased, and damaging" to religious views is a violation of the First Amendment: "The government is establishing, as the official religion of the United States, secular humanism." The U.S. District Court for the District of Columbia dismissed the case in May 1973, ruling that the First Amendment does not allow the state to require that teaching be tailored to particular religious beliefs. Willoughby then went to the U.S. Supreme Court, which dismissed the case in February 1975. In 1978, the creationists also went to a U.S. District Court to sue the Smithsonian Institution, petitioning the Court to block a museum exhibit called "The Emergence of Man."[18] The District Court's ruling against the creationists was upheld in 1980 by the U.S. Court of Appeals.

Undaunted, creationists again took the initiative in California, suing the California Board of Education in 1980 for violating the religious rights of children. The case was touted as a "rerun of the Scopes trial," "the trial of the century," and a "test of religious freedom." The plaintiffs lined up school children to plead that children should not be told that their religious beliefs are wrong. In defense, the State Attorney General mobilized eminent scientists as expert witnesses to vouch for the validity of evolution theory. However, during the trial, the creationists reduced their complaint to an administrative detail concerning the wording of the guidelines in the *Science Framework*. The judge recommended that the Board of Education circulate a policy statement emphasizing the need to reduce dogmatism—hardly a major victory for the creationists. But the event brought publicity and media attention and once again helped to legitimate the creationists' cause.

The 1981 Arkansas case must be viewed, then, in the context of a series of legislative actions and legal challenges to the constitutionality of creationist demands. In 1973, less than six years after repealing its anti-evolution legislation, the Tennessee legislature passed a new statute requiring that

Any biology textbook used for teaching in the public schools, which expresses an opinion of, or relates a theory about origins or creation of man and his world shall [give] . . . an equal amount of emphasis on . . . the Genesis account in the Bible.[19]

The National Association of Biology Teachers joined with Tennessee biology teachers and parents to challenge the constitutionality of the legislation, contending in Federal District Court that the law interfered with free speech, free exercise of religion, and freedom of the press as guaranteed by the First and Fourteenth Amendments.[20] Finally, in 1975, the U.S. Court of Appeals overruled the equal-time legislation, arguing that it showed

. . . a clearly defined preferential position for the Biblical version of creation as opposed to any account of the development of man based on scientific research and reasoning. For a state to seek to enforce such preference by law is to seek to accomplish the very establishment of religion which the First Amendment to the Constitution of the United States squarely forbids.[21]

Undeterred, the creationists introduced bills into the legislatures of some twenty other states. Learning from the failure in Tennessee, they sought to develop legislation that would not violate the First Amendment. A creationist lawyer, Wendell Bird, wrote a model resolution that he felt would "neutralize" the teaching of scientific Creationism by sharply distinguishing it from religious Creationism. This resolution was turned into the "Balanced Treatment of Creation Science and Evolution Science" Act which passed in Arkansas and later in Louisiana.

The plaintiffs, a coalition of parents, ministers, and community leaders represented by the ACLU, challenged the legislation on three grounds: that it violates the First Amendment because it is religious and not scientific in purpose and advances the religious beliefs of fundamentalists; that it abridges academic freedom of students and teachers by forcing the teaching of a doctrine which has no scientific merit; and that it is unconstitutionally vague, allowing unfettered discretion over its enforcement.[22] Philosophers, theologians, sociologists, scientists, and educators testified that—historically, philosophically, and sociologically—Creationism is a religious movement and that "creation-science" is no more than religious apologetics.

The defense of Creationism as a scientific model hinged on distinguishing between the notion of a creator and a deity, as expressed in a remarkable (and surely heretical) statement in the defendants' legal brief:

Assuming *arguendo* that a "creator" and "creation" are consistent with some religions, this does not make these inherently religious. The entity which caused the creation hypothesized in creation-science is far, far away from any conception of a god or deity. All that creation-science requires is that the entity which caused creation have power, intelligence, and a sense of design. There are no attributes of the personality generally associated with a deity, nor is there necessarily present in the creator any love, compassion, sense of justice, or concern for any individuals.[23]

Witnesses for the defense included seven scientists who tried to discredit evolution biology by showing what it could not explain; however, their testimony revealed the difficulties in trying to distinguish science from religion in order to meet constitutional imperatives. As witnesses tried to document the scientific nature of Creationism, their fundamentalist beliefs prevailed. Under cross-examination, several of the state's witnesses acknowledged that their guidance and, indeed, their evidence for creation came directly from the Book of Genesis. A key witness brought in to show the "enlightened and scientific attitude of today's creationists," sounded much like the 1920s' fundamentalists when he expressed his belief in demonic possession, exorcism, occultism, and UFOs as "a satanic manifestation in the world for the purpose of deception."[24]

Judge William R. Overton's decision in favor of the plaintiffs is sharply worded and unambiguous. In the statute, he finds "inescapable religiosity," a "hodgepodge of limited assertions many of which are incorrect," "continued dualism which has no scientific factual basis or legitimate educational purpose," "meaningless assertions," and "fallacious pedagogy." He argues that creation-science fails to conform to the essential and accepted characteristics of science: "It is not explanatory by reference to natural laws, is not testable, and is not falsifiable. . . . they take the literal wording of the book of Genesis and attempt to find scientific support for it." To comply, schools would have to "forgo significant portions of subjects such as biology, world history, geology, zoology, botany, psychology, anthropology, sociology, philosophy, physics, and chemistry." Concluding that the Act was "simply and purely an effort to introduce the Biblical version of creationism into the public school curriculum," Judge Overton issued a permanent injunction on the law.

The Arkansas state senator who had introduced the creationists' bill announced even before the decision that

. . . if we lose it won't matter that much. If the law is unconstitutional, it'll be because of something in the language that's wrong . . . So we'll just change the wording and try again with another bill . . . We got [sic] a lot of time. Eventually we'll get one that's constitutional.[25]

The creationist lawyer, Wendell Bird, dissociated himself from the case well before the decision. Others had also laid the groundwork for dismissing the case as biased and therefore meaningless as a test of the constitutionality of creationist legislation. During the trial, television preacher Pat Robertson and Moral Majority leader Jerry Falwell accused the Arkansas Attorney General of collusion with the ACLU. The Creation Science Legal Defense Fund also attacked the Attorney General for inadequate preparation. Roy McLaughlin, a Moral Majority spokesman, questioned Judge Overton's neutrality and called the decision a form of censorship. Duane Gish of the Institute for Creation Research characterized the decision as a blow to academic and religious freedom: "If anything, the creation-scientists' efforts will be intensified."

Indeed, before the trial was over, creationists were redrafting their model act in light of the ACLU challenge. They changed its name from "Balanced Treatment" to "The Unbiased Presentation of Creation-Science and Evolution-Science," because it is more difficult to argue against the removal of bias. And they modified any statements or references that could be construed as religious or supernatural. This new draft is now circulating in legislatures throughout the country.

Despite their failure in Arkansas, the creationists are encouraged by evidence of public support. In an astonishing editorial on Judge Overton's decision, *The Wall Street Journal* warned against using the decision to dismiss the "political discontent" that gave birth to the law:

If caught between the relativists and the fundamentalists, we ourselves might often be tempted to side with the fundamentalists, at least those who are concerned, as we are, about a decline in the moral order. . . . We are not sure the courts have considered what it may be like if they insist on divorcing government entirely from spiritual thought. . . . We hope that the forces who have won this narrow battle in court won't labor too long with the notion that they have scored some major victory against religious belief.[26]

This editorial assessment of public opinion is supported by a 1981 national public opinion survey conducted by NBC News: 76% of the Americans polled felt that public schools should teach both the scientific theory of evolution and the Biblical theory of creation, 8% favored teaching only the scientific theory, and 10% only the Biblical theory.[27] Just as in the 1920s, creationists are a dramatic expression of American society's renewed concern with moral and religious values; and their concern is again focused on science as a dominant world view that threatens traditional values.

Creationists are part of a broader social movement evident in the revival of Fundamentalism, the success of radio and television preachers, and the significant increase in political activism among groups with religious intent. Richard Viguerie, the direct mail expert for the "New Right," describes this activism as an "ideological war against the godless minority of treacherous individuals who have been permitted to formulate national policy."[28] The "war" has many goals—reinstating prayer in the schools, blocking the Equal Rights Amendment, prohibiting abortion, banning books, and removing the civil rights of homosexuals— but science education is a critical target because it represents cultural diversity and secular trends. In the fundamentalist tradition, New Right activists resist the influence of "secular humanism," a term loosely used to describe a world view that is human-centered and secular and that emphasizes the ability to achieve self-realization through the use of reason and scientific method, rather than through belief in a spiritual and moral order. For the Moral Majority, secular humanism is a "civil religion"—shorthand for the forces of evil.

The creationists are part of this revival of Fundamentalism, reflecting a quest for order and authority in a society increasingly influenced by the censors of the Radical Right. Backed by the support of the Right, creationists turn defeat to their advantage, using it to enhance their image as a beleaguered group that is rejected by a scientific establishment bent on protecting itself against "new" ideas. Using representations that are well-adapted to the twentieth century, and claiming scientific respectability, they offer intellectual plausibility as well as salvation, and the authority of science as well as the certainty of scripture. Because they have fused three venerated traditions of American culture—science, religion, and populist democracy—the creationists' influence is likely to persist.

References

1. Philip Appleman, ed., *Darwin* (New York: Norton, 1970); Charles Gillespie, *Genesis and Geology* (Cambridge, MA: Harvard University Press, 1951).

2. *State of Tennessee v. John Thomas Scopes*, Nos. 5231 and 5232 in the Circuit Court of Rhea County, Tennessee (1925). *Scopes v. State*, 154 Tenn. 105, 289 S.W. 363 (1927).

3. Arthur Garfield Hays, "The Scopes Trial," in *Evolution and Religion*, Gail Kennedy, ed. (New York: Heath, 1957), p. 36.

4. Chester H. Rowell, "The Cancer of Ignorance," *The Survey*, 1 November 1925.

5. Sheldon Norman Grebstein, ed., *Monkey Trial: The State of Tennessee v. John Thomas Scopes* (Boston, MA: Houghton Mifflin, 1960), p. 126.

6. S. J. Holmes, "Proposed Laws Against the Teaching of Evolution," 13 *Bulletin of the American Association of University Professors* 8 (December 1927): 549–554.

7. Hermann J. Muller, "One Hundred Years Without Darwinism Are Enough," 19 *The Humanist* (1959): 139.

8. W. Dykeman and J. Stokley, "Scopes and Evolution—The Jury Is Still Out," *The New York Times Magazine* (12 March 1971): 72–76.

9. Repeal was voted by the Tennessee Legislature in 1967. See Dorothy Nelkin, *The Creation Controversy: Science or Scriptures in the Schools* (New York: Norton, 1982).

10. *Epperson v. Arkansas*, 393 U.S. 97 (1968) struck down Initiated Act 1 of 1929, Arkansas Statutes Ann. Sections 80-1627-1628, which prohibited the teaching of the theory of evolution.

11. Creation Research Society brochure.

12. Richard Bube, "Science Teaching in California," *The Reformed Journal* (April 1973): 3–4.

13. Quoted in *Time* (16 March 1981): 82.

14. John Conlan, "The MACOS Controversy," *Social Education 39* (October 1975): 391.

15. California State Department of Education, *Science Framework for California Public Schools* (Sacramento: 1970), p. 106.

16. Vernon Grose, "Second Thoughts About Textbooks on Sexism," *Science and Scripture* 4 (January 1974): 14.

17. *William Willoughby v. H. Guyford Stever*, No. 15574–75 (Denver District Court, 18 May 1973), *affirmed*, 504 F.2d 271 (D.C. Circuit 1974), *certiorari denied*, 420 U.S. 927 (1975).

18. *Crowley v. Smithsonian Institution*, 462 F. Supp. 725. Case *affirmed* 636 F.2d 738.

19. Chapter 377 of the 1973 Public Acts of Tennessee, Amendment to the *Tennessee Code Annotated*, Section 49-2008, passed 30 April 1973.

20. *Daniel v. Waters*, 515 F.2d 485 (1975).

21. *Ibid.*, p. 489; also quoted in National Association of Biology Teachers, *News and Views* XIX (April 1975).

22. *McLean v. Arkansas*, U.S. District Court, Eastern District of Arkansas, Western Division, Civil Action LRC 81-322, Plaintiffs' Pre-Trial Brief filed 2 December 1981.

23. *McLean v. Arkansas*, U.S. District Court, Eastern District of Arkansas, Western Division, Civil Action LRC 81-322, Defendants' Pre-Trial Brief filed 2 December 1981, p. 15.

24. Author's notes on trial testimony.

25. Quoted in the *Washington Post* (7 December 1981).

26. Editorial, *The Wall Street Journal* (8 January 1982).

27. *The New York Times* (18 November 1981).

28. Richard Viguerie, *The New Right: We're Ready to Lead* (Ottawa, IL: Caroline House Publishers, Inc., 1981).

The Arkansas Creationism Trial: An Overview of the Legal and Scientific Issues

Eric Holtzman
and
David Klasfeld

How can a segment of knowledge be identified as "scientific"? Who qualifies as an expert in a science? Do the criteria put forth by philosophers of science aid in making such decisions? When is a scientific debate to be regarded as closed? When it is, what should become of minority opinions? These are just some of the scientific and philosophical issues that formed a backdrop to the proceedings at the Arkansas "Creationism" trial.

The Arkansas case was explicitly about the separation of church and state: in the decision, Creationism was recognized as a particular religious viewpoint and, as such, its promulgation via publicly funded schools is prohibited in the U.S. Constitution. Implicit in this case, however, are more general scientific and philosophical issues that surfaced in various guises and at various points in the trial, the planning of strategies, the selection of witnesses, and the taking of pre-trial testimony (depositions). The most fundamental such question is: Who should decide what is to be taught as *science* in the schools?

Our essay explores several such issues as they became refracted through the processes of planning the presentation of the plaintiffs' (anti-creationist) case and through the trial itself.

Creation "Science"

The creationists' claim, as articulated in the Arkansas law that they sponsored, was that Creationism has the same scientific status as evolution theory and is not inevitably a religious doctrine. Unlike the versions of creation found in other religions, the particular account of creation put forth in Genesis, they argue, is as defensible scientifically as is Darwinism and its offshoots. These assertions were necessary in the light of previous legal findings that teaching explicitly religious material violates the constitutional strictures on separation of church and state. Their transparency is, of course, self-evident. For instance, a number of leading creationist textbooks are written in two versions—one quoting the Bible as authority, the other making precisely the same arguments but omitting all explicit references to Scripture or God; also, members of the Creation Research Society (a focal organization in the struggle for legitimization) must not only have a degree in some science but also sign an affirmation of their belief in the literal truth of the Bible.

The creationists' self-imposed logical contortions lead them to make—or, more precisely, to resurrect*—a scientific and philosophical attack on the theory of evolution on three fronts. First, they cite disagreements among evolutionists as demonstrating that evolution theory is full of holes and inadequacies. Second, they argue that the theory of evolution, unlike the physical sciences, is not directly verifiable and hence cannot be proven by the accepted norms of science. Third, they argue that there are only two possible theories of "origins"—all others being reducible to one of the two—so that any evidence against evolution is evidence for their brand of Creationism. Creationism and evolution are thus placed on equal footing—one can cite evidence interpretable within the logic of either position but no decisive proof of either is possible within the framework of true science. The implication is that the creationists' procedure—the forcing of facts into an interpretative framework based on faith—is no different from the evolutionists' procedures.

Responses I

From one point of view, the creationists' arguments are easy to meet. Their texts and presentations include elementary errors, as well as distortions of fact. Even the most well-known of the defense's scientific witnesses at the Arkansas trial knows very little biology; once this was

*Many of the creationists' arguments, including their calculation of the likely age of the universe, echo positions taken by their counterparts a century ago in attacking Darwin himself.—EH/DK

brought out in his deposition,* he took pains to argue in his public testimony that such self-admitted ignorance confers the advantage of an open mind.

Creationism attacks not only the Darwinian theory of evolution, but also all areas of astronomy and geology predicated on the assumption that the universe is older than the 10,000 or so years the creationists claim as the likely age of the universe (based on genealogical information in the Bible). In order to conform to the creationists' ideas, the branches of physics dealing with thermodynamics and with radioactivity and the chemical theories underlying radioactive dating would also have to be revised. In other words, the creationist critique, if accepted, would demand wholesale revision of much of the science curriculum. As was brought out at the trial, revising the accepted theories or science would, for example, even complicate teaching petroleum geologists how to look for oil—in short, the stakes for science students extend beyond abstract knowledge.

Creationism also cannot provide positive evidence for creation. The thrust of the creationists' scientific case is negative—they point to gaps in the fossil record or to scientists' disagreements about evolutionary mechanisms, for example. This strategy, however, puts them at a severe disadvantage because scientists have regarded most of the debate on which they focus as having been settled for almost a century, and the few new issues the creationists emphasize were originally uncovered and are enthusiastically investigated by evolutionists. In the Arkansas trial, the creationists did try to demonstrate, through testimony by scientists sympathetic to Creationism, that their point of view leads to alternative scientific hypotheses and orientations that are proving fruitful in research. But this claim's strength is largely vitiated by the quality of the product. Although the findings presented at the trial and in the creationist literature do occasionally include an interesting or puzzling fact, none of the empirical work necessitates revision of conventional thinking.

The creationists neither understand science, nor see its overwhelming coherence and momentum. They are scientists wearing blinders, picking and choosing among all the facts to find only those few which provide

---

*A "deposition" is a pre-trial proceeding in which potential witnesses, under oath, must answer questions posed by the attorneys for both sides. The term may also be used to refer to the testimony so taken.—Ed.

some support for their position and ignoring all contrary material, no matter how fundamental.

Should Evolutionists Shut Up?

Even though the creationists' particular case can be easily refuted, this does not exhaust interest in the issues they raise.

One question is the extent to which evolutionists are responsible for fueling the creationists' attack. There is, no doubt, a common tendency among evolutionists to overstate their internal disagreements and to dramatize individual positions in a scientific debate by stressing the challenge a position may pose to conventional wisdom. Biologists are no more modest than any other academics in assessing the uniqueness of their own thinking.

In the past few years, paleontologists and geneticists have engaged in luxuriant debates over the nature and pace of evolutionary change. The British Museum, for example, in designing recent exhibits on evolution, included in the material presented some vague statements interpretable as condoning all sorts of heterodox views of evolution, even creationist ones. These statements triggered a lively exchange in *Nature* among evolutionists, theologians, and many others.

The fall-out from such a debate is manna for the creationists because their style is casuistical and their core of evidence consists partially of quotations from evolutionists bemoaning their own ignorance or that of "misguided" colleagues. In response, there have been calls for "responsibility" in public debate among scientists and for a pulling together of the community of evolutionists to confront a dangerous obscurantist challenge. The tangle of problems that would arise from the suggested self-censorship, and the defeat of highly regarded principles such censorship would represent, are obvious. In fact, the Arkansas proceedings prompted among the scientists an undertone of "tragedy," focusing on the diversion of time and effort from scientific exploration to self-defense against an opponent whose ultimate interests have little to do with the pursuit of scientific knowledge. Still, open questions about the responsibilities of scientists to the public and about the promulgation of mass scientific literacy cannot easily be evaded. The public television "Life on Earth" series, for example, dazzling as it is, cannot penetrate a mass audience broadly enough to compete with the "pray-for-pay" programs on the religious networks.

Responses II: Lawyers and Scientists

In developing the case, attorneys for the plaintiffs sought advice from over one hundred scientists, historians, philosophers, and other social scientists. Given the nature of the academic environment, these consultants were almost as accustomed to combative modes of intellectual efforts as were the lawyers. They were also accustomed to cooperative team endeavors—the sharing of strengths. But divergences in customary audiences and in the way in which evidence is employed contributed to some ongoing debates among the groups developing the case for the plaintiffs.

Not surprisingly, the lawyers focused their concern on constructing a case that dealt with the specific issues before the court. The lawyers first had to assimilate, understand, and synthesize the evidence and then put it in a form and style appropriate to the non-expert (in this case, the judge). Although the judge was expected to reach a decision that in the final analysis had to be "yes" or "no," it was completely within the judge's power to decide the case either on a narrow or on a broad basis, on technicalities or minor points, or on fundamental constitutional and legal grounds. The case therefore could not dwell excessively on the open-ended facets of science and philosophy, but had to be argued on conclusive points.

In contrast to the lawyers, the scientists were accustomed to marshalling evidence in areas of their own direct expertise for the scrutiny of other experts, to leisurely, fine-grained analysis made possible by publication and "free-wheeling" debate, and to the substitution, when necessary, of provisional decisions for final ones. At the trial, the scientists were somewhat impatient to counter the creationists' more outrageous claims, even if such refutation did not directly contribute to the case.

As might be expected, the philosophers had the most diverse responses. Some reveled in the ambiguities that arise in trying to define science (after all, Karl Popper himself has publicly reversed a portion of his initially negative position on the extent to which evolution theory is scientific). Others savored the rare opportunity to use philosophical tools in "real world" struggles.

How to Focus the Case?

Both the lawyers and the scientists first had to undergo a form of re-education. When approached for assistance, scientists frequently seemed to believe that sending a few experts to Arkansas to inform the judge that creation "science" was nonsense and to show how flimsy the creationists' arguments are would be sufficient to stop the "silliness" then and there, ignoring (at least initially) that a state legislature had passed the law and that the "creation-science" movement was national in scope.

The lawyers thought that it must be easy enough to define "science" and then to demonstrate that evolution is, and creation-science is not, science. They thought that this sequence would provide an easy-to-comprehend litmus test that would also steer the testimony away from technical discussions on the many scientific matters for which the creationists have trumped up arguments.

How then to construct an effective case that could maintain intellectual integrity? Should there, for example, be a point-by-point refutation of the creationists' "science"—the facts and arguments with which they attack evolution theory? Although this strategy would be easy for any given point, it could leave the impression of a contest between positions of comparable weight. In addition, the creationists are adept at public debate in which they stay on the offensive, diverting attention from the fissures in their arguments by highlighting each flaw, no matter how minor, in opposing views. They skip from one matter to another and shift ground rapidly from biology to astronomy, to geology and back again, and are entirely unabashed when shown to be inaccurate or wrong. A comparison of the negative aspects of the contending positions, therefore, was inappropriate as a foundation for the case.

The plaintiffs' scientists pointed out that a strong positive case could be made because evolution theory is, after all, one of the central accomplishments of biology and an important underpinning for advances in modern biology. Genetics and molecular biology have achieved an extraordinary deepening and an expansion of evolution theory, revealing underlying mechanisms that Darwin could not know about and which the creationists distort or ignore in the parody of evolution that they attack. Furthermore, as already mentioned, biology, geology, cosmology, chemistry, and other sciences are interlinked in their contributions to the study and theory of evolution.

This approach shows the proper status of the gaps, flaws, and disagreements stressed by the creationists as reflecting the strengths rather than the weaknesses of the sciences. Witnesses for the plaintiffs could readily discuss the inadequacies of present knowledge within the framework of understanding that science thrives on the search for anomalies and that scientists make their careers in seeking explanations for what is not known. The picture of evolution theory as a living science, carried on by disputatious people who nonetheless start from common basic premises contrasts sharply with the one the creationists' "science" witnesses attempted to convey.

The lawyers felt the focus should be on creation "science," not evolution. To focus on evolution, they believed, no matter how important the science, would allow the creationists to fight on friendly terrain— i.e., the perceived "failings" of evolution theory. The lawyers wanted the creationists also to have to defend less familiar territory—e.g., positive assertions about creation science.

The ultimate presentation for the plaintiffs was an amalgam: an attack on creation-science by scientists who at the same time could testify about the enormous scientific advances since Darwin which have supported or extended his theory.

In the defense case, the creationists made long, idiosyncratic presentations of their individual interpretations of minor scientific findings. The presentations had clearly been styled for a missionary setting; arguments were based on "common sense" to convert the unconverted. The creationist witnesses seemed to imply that a biased scientific community ignored their views and findings; but, on cross-examination, the creationists were constrained to admit either that they had not submitted their findings to the usual scrutiny of editorial boards and other peer groups or that they had not in fact had their findings discussed or published at all. Many of the creationists have degrees in science, but they frequently are arguing well outside their areas of expertise. Their findings, in general, are irrelevant and tend to be uninteresting or, more accurately, to be minor or readily explainable within the conventional scientific framework.

When, for example, we took the deposition of a creationist mycologist (a tenured professor at a major state university), he testified that he had been studying the relationship between hosts and parasites for fifteen years and in his opinion that relationship was so sensitive and so complex that the hosts and parasites must have been created in exactly that relationship because they could not exist if there were any alteration in the relationship. He also testified that he had never discussed this theory with any other expert in the field, never published an article on his theory, knew of no other mycologist who shared his views, and had done no experimental work that supported his theory.

What is Science?

The trial largely bypassed the deeper issues about "what is a theory," "what is science," and "who is a scientist." Science was instead presented as what the recognized scientific community treats as such. This approach does have troublesome facets as a general position, especially if the scope also includes fields of study other than the natural sciences, where the credentials of experts are not as "unambiguous" as they are in the natural sciences and where a modest technical background seems unessential for evaluating the conflicting claims. Still, it does not seem inappropriate to maintain that the primary or secondary school classroom is not the place where challenges by small insurgencies to a virtually unanimous scientific community are first to be decided. The creationists' "free-speech" arguments—"let the students decide"—have an intuitive appeal. But this is lost when one thinks about the actual relative status and power of science and religion in everyday life and remembers that the issue at stake is simply what is to be taught in a *science* classroom, not whether students should ever be told that there are religious viewpoints to which some people hold strongly.

The strategies adopted in the plaintiffs' case developed in part as a result of the plaintiffs' attorneys' initial efforts to develop criteria that could be presented explicitly as distinguishing science from nonscience or good science from bad science. It soon became clear that this effort was foredoomed as an exclusive argument. Tempting as it is to say that evolutionists are good scientists and creationists are not, developing such a position as the centerpiece of a public trial is not as easy as it looks. Creationism was, after all, not such bad science a century ago, and one can easily cobble an argument that Creationism still has scientific aspects and does stimulate the evolutionists to try a little harder. More importantly, the philosophers of science have spent many decades poking holes in each other's view of what science is and is not and there certainly is no set of universally acceptable criteria. The conventional testability/verifiability/falsifiability doctrines apply poorly to the old style of evolution theory—whence Popper's initial doubts. These doctrines do much better with the modern genetic and molecular phases of the theory. But on the topic of initial origins of life on earth, accusations of irremediable speculativeness and irreproducibility are made. Current theories of evolution do not require any particular view of the origins of life—they concern change in living forms. However, the issue of the origins of life is an emotionally powerful one that inevitably arises forcefully in the creationist attack and is specifically referred to in the Arkansas statute. Even the creationist scientists' self-damaging testimony that they base their thinking on acceptance of the Bible can be counteracted, crudely or eruditely, according to one's taste, through the by now commonplace conclusion that all scientific thinking starts from preconceptions. In other words, distinguishing science from non-science on abstract grounds is not all that easy.

In the final analysis, one is either left with all the ambiguities dangling or driven to adopt the criteria actually used: even though the rules adopted by the scientific community to govern its own activities are hard to codify, change with time, and apply in different ways to different sciences, in the most general terms those practices that encourage free rein to ideas and experiments are "good" science and those that defend rigid, formal preconceptions (Genesis, for example) tend not to be. Evolutionists, even if often blind and biased, are *sometimes* driven by the nature of their work to question all of those preconceptions of which they are aware, whereas creationists start by taking their central, conscious preconception as *necessarily true* and controlling. This is the essence of the claim that creation-science is religious and evolution theory is not. One can cast this distinction in terms of testability and so forth, and to a limited extent this was done at the trial. But to insist that the judge attend to and render judgment exclusively on the philosophical niceties seemed a pointless exercise in self-indulgence. Even more dangerous—from the viewpoint of the principles of academic freedom and healthy scientific practice—would be to suggest that a judge decide how good evolution science is or how bad the science in creation-science is. Therefore, the essential thrust of the plaintiffs' effort was to demonstrate that, at its heart, creation-science is not science at all but rather is thinly disguised religion. A few examples of creationist "science"—such as the use of the impossibly vague Biblical term "kind" to classify the different sorts of organisms—sufficed to demonstrate the consequences of creationist closed-mindedness for both scientific practice and education.

Evolution from Space

The selection of Chandra Wickramasinghe as a witness by the defense exemplifies the effects of the overall creationist tendency to see only those facts that are helpful to their argument and ignore those that are not. Because he and his senior collaborator, astronomer Fred Hoyle, are well-known scientists who challenge the conventional theory of evolution, Wickramasinghe was called to support the creationist case. In fact, he did it great damage.

On the one hand, Wickramasinghe was compelled to admit that he had virtually no training in biology, being a mathematician and astronomer by trade; but he felt free to advance a detailed theory of evolution based on a few scattered astronomical facts, which he presented at length, and a great deal of speculation about genes and microbes carried to earth by comets (the latter proposals resembling elements of Fred Hoyle's novels). On the other hand, he does know enough science to recognize the flaws in Creationism and thus he testified "that no rational scientist could believe that the earth was less than one million years old" and "that the earth's geology could be explained by one worldwide flood," both of which are cardinal tenets of the creationist position.

Wickramasinghe's testimony also brought out the fact that one can conceive of many variants of Creationism, not just the one in Genesis, thereby damaging the defense's assertion that there really are only two theories of "origins."

We first questioned Wickramasinghe at a deposition. This is a legal procedure designed to minimize the element of surprise in courtroom testimony, because questioning in the deposition can be wide ranging. The witness also cannot subsequently change her or his testimony because the answers given at the deposition can be and are cited during the trial itself.

As did many of the other creationist witnesses, Wickramasinghe enthusiastically expounded his theory at the deposition, seeming particularly anxious to impress the scientists present as advisors.* In doing so, he demonstrated startling ignorance of rudimentary biology.

*As in each of the depositions of the defendants' scientists, there were advisors present who had expertise in the areas of anticipated testimony. At the Wickramasinghe deposition, there were an astronomer, two biologists, and a paleontologist; these advisors suggested lines of inquiry and specific questions to the lawyer (who is the only one permitted to ask questions at a deposition).— EH/DK

The scientists advising the plaintiffs wanted Wickramasinghe's scientific theories to be taken apart publicly, but the lawyers decided to concentrate on demonstrating Wickramasinghe's own profound disagreements with the creationists' "science." Judge Overton's remark that he wondered why the defense had called Wickramasinghe as a witness, since he disagreed with most of their basic theory, indicates that the latter strategy was successful.

To Take Home
Perhaps the most dramatic aspect of the Arkansas proceeding was the weakness of the creationist case, a matter not of presentation or preparation, but of the fundamentals. As the final ruling implies, the Creationism movement, in its legal and legislative phases, is caught in the unpleasant situation of having to find gimmicks rather than being able to demand openly the teaching of religious doctrine. Those who oppose the present fundamentalist thrust for social control are taking solace in the strength and breadth of Judge Overton's decision; but, at its heart, Creationism is part of a widespread political movement that plays on some basic prejudices, on some genuine and justified resentments, and on the peculiarities of American religious and intellectual traditions. Thus, the creationists certainly will not accept the spirit of any court decision, even if they are constrained to accept the letter. The battle will continue at the level of school boards and textbook publishing houses for some time. What the Arkansas case does is to serve unequivocal public notice that the goal of Creationism continues to be to force evolution out of the schools, not to construct legitimate scientific alternatives.

Trying Creation: Scientific Disputes and Legal Strategies

Mark E. Herlihy

The "creation-science" controversy, which smouldered in this country (particularly in California) for more than twenty years, has again flared into prominence. The creationists' efforts to secure through legislation a legitimacy denied them in the scientific and educational communities have shifted the milieu of the controversy to the Federal courts, and away from the less coercive (although arguably more suitable) fora within specialized communities possessing the expertise to evaluate the creationists' claims. The first round of these contests resulted in a definitive legal ruling declaring Act 590, the Arkansas "creation-science" statute, unconstitutional. But the case also raised numerous questions regarding the role of the courts in such disputes, and the interpretation to be attached to judicial pronouncements on such matters as the definition of "science" or "religion." An adequate treatment of such questions can only be afforded by explicating the legal significance of some of the apparently non-legal elements of Judge Overton's opinion, and in the process exposing some of the features of the strategies employed in the litigation of such issues. This paper is intended to provide, in a preliminary fashion, such an explication from the point of view of a legal practitioner advancing plaintiffs' arguments that Act 590 was unconstitutional.

The central legal issue presented in *McLean v. Arkansas*[1] was never in any real doubt. Since the seminal case interpreting the Establishment Clause of the First Amendment (*Everson v. Board of Education*[2]), it has been the law that, to survive constitutional challenge, legislation must have a secular purpose and must have a "primary effect" that neither advances nor inhibits religion. In addition, the legislation must not result in the excessive entanglement of government with religion (*Lemon v. Kurtzman*[3]).

The Arkansas statute, however, was based upon a model act drafted to avoid the First Amendment's strictures. Thus, Act 590 on its face purported to foreclose explicitly "religious" instruction by deleting references to God or scriptures, and purported to establish a "secular" purpose by promoting the "academic freedom" of students and parents who found "humanistic, evolutionist" theory offensive.

The drafters of the model "creation-science" act also sought to insulate it from judicial review by the device of including in the text "legislative findings of fact" intended to establish defenses to a constitutional challenge. Relying on the general doctrine of separation of powers, courts have been reluctant either to disregard or to review factual determinations made by the legislative branch when properly executing its legislative responsibilities. Although the doctrine of separation of powers is strictly applicable only between co-equal branches of government (e.g., Congress and the Federal judiciary), the doctrine of state-Federal "comity" lately in vogue in Federal jurisprudence suggests that similar deference be paid by a Federal court to a state legislature.

The drafting of the creationists' model act thus had two objectives: first, to create a facially constitutional statutory text, and second, to insulate the contents of that text as much as possible from judicial scrutiny. Conversely, the task for those seeking to challenge Act 590 was, first, to broaden the inquiry by the Court into the Act and its antecedents and, second, to pierce the text of the Act and demonstrate, through various analyses aptly described as hermeneutic, that the seemingly secular statutory language cloaked a fundamentalist apologetic. This task was complicated by what might in the context be called the burden of persuasion borne by Act 590's challengers. It was not adequate to demonstrate that the statute was "merely consistent" with the interpretation offered; rather, it was necessary to establish an intentional connection between the text and the broad religious context in which the Act's challengers wished it viewed.

Obviously, such a challenge did not contemplate thrusting a Federal judge into the role of an arbiter of scientific claims, religious claims, or sociological descriptions. It did attempt as much as possible to permit the judge to remain in the familiar position of an interpreter of a statute in light of its language, legislative history, social context, and likely effect. Consistent with this approach, the expert testimony at trial was intended to offer the trial judge an understanding of the history and social context of the "creation-science" movement, of the consideration and conclusions of the scientific and philosophical communities regarding the status of "creation-science" as science, of the relationship between the "two-model approach" enshrined in the Act and the history and theology of Christian Fundamentalism, and of the impact of Act 590 on the educational system within Arkansas. This expert testimony was the critical element of the constitutional challenge to Act 590. A proper understanding of Judge Overton's use of that testimony in his opinion is central to understanding of the opinion and its implications.

In general, the role of an expert at trial is distinguished from that of an ordinary witness by the sole fact that the expert, having first established qualifications to do so, is entitled to opine on issues, primarily of fact, before the court. This ability to render opinions is both an aid and a challenge to a judge. It provides the court with a means of determining legally significant facts inaccessible to a judge without particular knowledge or experience in the expert's field. But it is a challenge to the extent that the expert usurps the judge's role by opining regarding a fact of determinative legal significance. Thus, for example, if the liability of an automobile manufacturer for injuries caused by an exploding fuel tank could be established through proof that (1) use of an unsuitable material in the tank construction caused the explosion, and (2) a competent automotive design engineer would or should know that fact, then an expert opining as to the truth of those facts will, in large part, have removed the question of liability from the judge, who can choose independently only to believe or not believe the expert. Judges are acutely aware of this potential influence and, especially in the face of conflicting expert testimony presented on opposing sides of an issue, frequently seek a basis for decision that avoids or reduces the significance of the expert testimony.

The issues presented in *McLean*, however, occasioned an interesting variation on the usual role of the expert. Particularly in the case of the science and education experts, the opinions and conclusions offered on behalf of the plaintiffs were not the opinions of experts applying their experience to a particular set of facts significant to the legal issues at trial, nor were they those of experts testifying about the absolute truth or falsity of any particular scientific fact or conclusion regarding the origin or development of the universe. Rather, those experts presented to the court the results of twenty years of the scientific and educational communities' evaluation of the scientific and educational claims of the creationists, set against one hundred years of experience since the Darwinian revolution.

This use of experts achieved two principal objectives. First, the trial judge was not asked to determine any scientific question of fact. For example, Judge Overton's opinion avoids judicial determination of the validity of even such a readily accepted scientific finding as the age of the earth. When discussing scientific findings, Judge Overton distances himself from the findings, avoiding their adoption in legal terms. For example, when discussing "the sufficiency of mutation and natural selection . . ." to account for speciation, an element of Act 590's "evolution-science model," Judge Overton finds the Act's assertion to be inaccurate because "Drs. Ayala and Gould . . . testified to such phenomena as recombination, the founder effect, genetic drift and the theory of punctuated equilibrium, which are *believed* to play important evolutionary roles."[4] Similarly, in discussing defense evidence, Judge Overton notes:

Robert Gentry's discovery of radioactive polonium haloes in granite and coalified woods is, perhaps, the most recent scientific work which the creationists use as an argument for a "relatively recent inception" of the earth and a "worldwide flood." . . . The discoveries have not, however, led to the formulation of any scientific hypothesis or theory which would explain a relatively recent inception of the earth or a worldwide flood. Gentry's discovery has been treated as a minor mystery which will eventually be explained.[5]

Such a finding explicitly relies on the state of affairs described by the expert, and can be readily distinguished from a fact found by the judge.

The second objective achieved by the type of expert testimony presented by the plaintiffs in *McLean* was the assertion in the legal forum of the legitimate, and indeed superior, claims of the specialized communities of scientists and educators to determine issues bearing directly upon their particular competence. As Thomas Kuhn has noted concerning the community of scientists,

One of the strongest, if still unwritten, rules of scientific life is the prohibition of appeals to heads of state or to the populace at large in matters scientific. Recognition of the existence of a uniquely competent professional group and acceptance of its role as the exclusive arbiter of professional achievement has further implications. The group's members, as individuals and by virtue of their shared training and experience, must be seen as the sole possessors of the rules of the game or of some equivalent basis for unequivocal judgments.[6]

By relying on the judgments of professional communities of scientists and educators, and by prudently avoiding making independent judicial determinations of scientific issues, Judge Overton remanded the scientific aspects of the Creationism debate to its proper forum, and neutralized the creationists' major tactic of appealing to the political process for legitimacy.

The most controversial element of the *McLean* opinion may be Judge Overton's use of the testimony of philosophical and scientific experts regarding the status of "creation-science" as science. Having avoided making scientific judgments, Judge Overton at first glance appears to have adopted judicially a particular philosophical view regarding the definition of science, and in so doing to have pronounced with juducial certainty upon an issue that in the philosophical community remains in substantial doubt. A closer examination, however, reveals that he is, in fact, behaving in a manner consistent with his treatment of scientific issues. He has relied upon philosophical testimony not to grant legal status to a definition of science but to articulate a distinction which is not in doubt between ordinary scientific endeavors and the discourse of creationists.

Viewed not as a normative definition of science but as a rule of distinction between scientific discourse and the discourse of creationists, the presentation to Judge Overton of the traditional logico-deductive model of science was eminently justified. In this context, it was not the role of expert witnesses to instruct the Court in the niceties of abstract philosophical debate but to provide the Court with a tool suitable to the task at hand. It is apparent from the creationists' own writings, as Judge Overton correctly points out, that creationists appeal to supernatural explanations of physical events, that creationists insulate their viewpoints from the rigors of empirical verification, and ultimately that creationists adopt a world view predicated upon a faith that rejects modification influenced by the appearances of the material world. Whether or not indices of testability, falsifiability, or mutability are appropriate as elements of an abstract philosophical definition of science, they are familiar elements of any self-referential discourse regarding science. Thus, if the *McLean* opinion adopts the view that science is what scientists do and that a scientific community exists, marked by its own rules, then reliance on this model of science cannot legitimately be faulted.

It is important to note that Judge Overton's treatment of scientific issues is not an independent element of his opinion. It is, in fact, embedded in a discussion of the text of Act 590. The core of Act 590, and of the creationists' rhetorical position, is the so-called "two-model" approach—which seeks to confine the universe of discourse about origins to a set of two mutually exclusive models, one of creation, the other of evolution. In rejecting the validity of such an approach (and more importantly in identifying such a restrictive view as characteristic of Christian Fundamentalism), Judge Overton again relied upon the testimony of experts, as well as his own common sense.

The testimony of religious experts established that Fundamentalism, historically and theologically, is based upon a peculiarly dualistic view of the world. The Arkansas legislature's adoption of such a dualistic view of the issue of origins could itself be viewed as a state endorsement of a sectarian religious view because this peculiar and distinctive dualism was common to both Act 590 and creation fundamentalists' religious beliefs. With the added fact that scholars in the field of Biblical hermeneutics could identify the substantive elements constituting the opposing models in the Act's text with the critical structural elements of the Genesis account of creation, there could be no doubt that the statute's purpose and effect was to advance religion.

In this context, Judge Overton's discussion of the scientific and educational status of "creation-science" was explicitly a negative treatment. Judge Overton did not, in the course of rejecting Act 590's establishment of fundamentalist religion, err by establishing contrary doctrines having to do with the definition of science or secular humanism. Judge Overton simply rejected the defensive claims of the creationists.

In conclusion, it should be recognized that the Federal courts, in resolving a controversy with such broad ramifications as that presented by Creationism, have a necessarily limited charter. In applying the Establishment Clause of the Constitution, a court may, as in *McLean*, find itself dealing with issues extending far beyond those that must be resolved judicially to fulfill the constitutional mandate of separation of church and state. The greatest success of the *McLean* opinion was its recognition of appropriate limits to the judicial resolution of disputes, and its implicit reliance upon the efficacy and integrity of specialized communities—such as the scientific and educational communities—to be the proper arbiters of claims of legitimacy made within their proper spheres.

References

1. *McLean v. Arkansas Board of Education*, 529 F.Supp. 1255 (E.D. Ark. 1982).

2. *Everson v. Board of Education*, 330 U.S. 1 (1947).

3. *Lemon v. Kurtzman*, 403 U.S. 602 (1971).

4. *McLean v. Arkansas* at 1267.

5. *Idem* at 1270.

6. Thomas S. Kuhn, *The Structure of Scientific Revolutions*, 2nd ed. (Chicago: University of Chicago Press, 1971), p. 168.

Science as an Apologetic Tool for Biblical Literalists   Gary E. Crawford

Advocates of legislation mandating "balanced treatment" for "creation-science" bristle at the suggestion that they are attempting to introduce religion into the public schools. They argue that creation-science can be presented solely in terms of "scientific" evidence and "inferences" from that evidence. Opponents of the legislation retort, of course, that creation-science is not science.

When "balanced treatment" advocates say creation-science can be presented without religious concepts, they mean that it is possible to eliminate the Biblical references and religious allusions now pervading their texts and to present only the empirical claims of creationists. These advocates say they can separate the physical from the meta-physical or, as a latter-day William Jennings Bryan might say, the rock from the Rock of Ages. These segregated "facts" they call "science."

Despite the admissions of several prominent creation-scientists that creation-science is not scientific theory,[1] and the recent judicial determination to the same effect in *McLean v. Arkansas Board of Education*,[2] the question continues to trouble a few non-creationist philosophers of science.[3] They argue that no sharp demarcation exists between what is scientific and what is not scientific, that the dividing line has continued to change throughout the history of science, and that there is no clear consensus among philosophers today on a definition of science. They also argue that, at least under some definitions, creation-science may qualify as scientific, although its testable conclusions have repeatedly been shown to be false and it has no secular educational value in the modern science class.

The encouragement "balanced treatment" advocates obtain from any suggestion that creation-science can be considered "scientific" and the consternation the suggestion causes to opponents of the legislation

are greatly disproportionate to the number of philosophers openly expressing concern. The dissenters are accused of giving aid and comfort
to the enemies of modern science (albeit unintentionally). The dissenters,
in turn, accuse philosophers who insist that creation-science is not
science of misrepresenting what science is and how it works because
of an overzealous resolve that Creationism must be discredited at any
cost.[4]

Partisans on both sides of the "balanced treatment" controversy take
the philosophical debate so seriously because many of them mistakenly
believe that its outcome is dispositive of the legal claim that the legislation is unconstitutional under the Establishment Clause of the First
Amendment. They reason that creation-science must be either "religion"
or "science" because it cannot be both and that "religion" is proscribed
by the Establishment Clause but "science" is not.

Such a simplistic, definitional approach to the Establishment Clause,
however, is not an adequate response to this sophisticated attempt to
circumvent the prohibition against government sponsorship or promotion of religious doctrine, and will not be adequate to meet similar
attempts likely to occur in the future. The questions necessary to determine what science is do not automatically identify the considerations
relevant in judging whether the educational process has been manip-
ulated for extra-scientific and religious ends—which is the critical inquiry
under the Establishment Clause. To fall within the proscription of the
First Amendment, creation-science need not be found to *be* a religion,
only to *advance* religion in such a direct way as to implicate First
Amendment values. The determinative constitutional question is not
into which philosophical classification creation-science fits but whether
the function (i.e., the purpose and effect) of teaching creation-science
in the public schools is to promote belief in sectarian doctrine and
thereby render the schools partisan to particular religions.

A functional analysis of creation-science under the Establishment
Clause focuses on whether the empirical claims of creation-science have
been developed through the ordinary standards of scientific discourse
or whether they are specially prepared as a polemical device for chan-
neling school children toward certain religious conclusions. Included
in this analysis is consideration of the following questions:

1. Do unarticulated religious criteria govern the selection, organization, or interpretation of the data?

2. Is the body of information intended to encourage acceptance of religious ideas?
3. Does the body of information have substantial support in secular scholarship?

Application of a functional analysis to creation-science reveals that it was developed and is disseminated to promote a belief that the book of Genesis, literally interpreted, is a scientifically and historically accurate statement of events and hence to engender a Biblical-literalist world-view. Such an analysis also shows that, as an integral theory, creation-science has no significant support in secular scholarship. Consequently, even if these empirical claims are assumed to fall within some abstract definition of science, it is clearly a science that is in the employ of religion, an apologetic tool constructed to encourage religious conversion.

Question 1: Religious Restrictions on the Evaluation of Data

The creationists' fundamental premise is that God has revealed Himself to human beings through the Bible, the "Word of God."[5] This revelation is absolutely authoritative for all aspects of human life. It is not allegorical, or symbolic, or beyond human comprehension, but is direct and straightforward, particularly in the first three books of Genesis, the story of God's creation of the world.

Several conclusions flow from this premise. Because the ultimate religious reality is in the God spoken of in the Bible and this God has given a clear account of how the world was created, human responsibility is to believe this account whether it contradicts science or history or any other human discipline. However, the creationists recognize that any appearance of conflict between science and religion in a society that values science as much as ours could be a serious threat to religious faith. To eliminate this conflict, creation-science seeks to "realign" empirical data within a framework of Biblical revelation. Creation-science, therefore, is an exercise in selecting, organizing, and, most importantly, interpreting facts observable in nature in such a way as to make them consistent with a literalistic reading of Genesis. The theoretical construct used to analyze information is perhaps most clearly articulated in Christian Heritage College's statement of educational philosophy. The Institute for Creation Research (ICR), the leading creation-science research and publishing center, was founded as part of the College and until recently was formally affiliated with it. Many of ICR's professional staff are adjunct professors at the College, including Henry Morris, the "father of creation-science," and Duane Gish, its most prominent spokesperson. The College asserts that placing the "real facts of science and history" in a "consistently creationist and Biblical framework" is "a highly effective mechanism for evaluating 'would-be' truths":

That is, in addition to the accepted secular criteria for testing such "would-be" truths (e.g., . . . scientific method, . . . human empirical experience), there is the infallible framework of truth revealed in Scripture by the Creator of all truth. Acceptance and application of this creationist philosophy and the basic Biblical criteria will thus optimize the search for truth and the correction of error. . . . [T]he ultimacy of truth as resident in the Godhead (John 14: 6; 15: 26) and in His Word (John 17: 17) does provide a necessary and sufficient criterion for testing and evaluating all alleged "truths" proposed through human research and reason. . . . [E]volutionary philosophy . . . should be thoroughly analyzed in terms of its supposed evidences and implications, even though the Biblical criteria require its rejection as possible truth.[6]

This means of evaluating science is illustrated repeatedly throughout the creation-science literature. For example, Henry Morris observed in a letter to the *Creation Research Society Quarterly*, "While, as scientists, creationists must study as objectively as possible the actual data of geology, as Bible-believing Christians, we must also insist that these be correlated within the framework of Biblical revelation. This restriction requires rejection of the traditional uniformitarian approach [of geology]. . . ."[7] In *Scientific Creationism*, Morris shows us this process of correlation:

The great Flood of Genesis 6–9 is of critical importance to the true understanding of earth history. It has been seen that sound Biblical exegesis will not permit placing the geological ages either before or during the six days of creation. Neither can the six days of creation be interpreted as non-historical or allegorical. The only other alternative is to reject the standard system of geological ages altogether.

This is of course a drastic suggestion—orthodox geologists indeed reject it out of hand. However, there is no other alternative. If the Bible is the Word of God—and it is—and if Jesus Christ is the infallible and omniscient Creator—and He is—then it must be firmly believed that the world and all things in it were created in six natural days and that the long geological ages of evolutionary history never really took place at all.[8]

Creationists begin with the "facts" divinely repeated in the Bible and select, describe, and interpret data from geology, paleontology, and other disciplines in such a way as to make them appear consistent with

Biblical revelation. In an attempt to disguise the religious restrictions on the scope of their scientific inquiry and thus, they hope, to make their texts suitable for use in public schools, creation-scientists eliminate from the public school editions any explicit mention of a Biblical framework for the interpretation of the data. The paragraph quoted above, for example, was removed from the Public School Edition of Morris' *Scientific Creationism*. As shown in Table 1, similar editing was done on Gish's *Evolution? The Fossils Say No!*. Contrary to the expectation of creationist editors, removing explicit religious references from creationist texts does not make them constitutionally more acceptable, only less honest.

Question 2: The Sectarian Objectives of Creation-Science

The literal truth of the Bible is important to creation-scientists not simply because it tells about God creating the world, but because of its message of salvation about "the Lord Jesus Christ." To the creationists, "Jesus Christ" is, in fact, the actual Creator. Creation and salvation are integral parts of the same theme: (1) initial perfect creation; (2) human sinfulness (Adam's Fall) resulting in (3) God's curse upon the world, leading to death and decay (which is identified with the second law of thermodynamics); (4) God's judgment upon the sinful world evidenced in the Flood (hence the importance of proving the Flood); and finally, (5) God's redemption of sinful humanity through the death and resurrection of "His Son," "The Lord Jesus Christ."

The centrality of the salvation message in creationist belief leads to a radical sectarianism. Creationists interpret the world around them in light of this salvation story and thus take an extreme, dualistic view of themselves and the outside world, centered around belief in creation or evolution. Belief in creation (and, by extension, salvation in Jesus) is identified with the only proper religious position. In contrast, evolution and belief in evolution are linked with all "anti-God and anti-Christian" belief systems. Given this interpretive framework, the creation-scientists' purpose is to save people from the pernicious effects of evolutionary thought.

Such a view assumes that Creationism's religious position is the only valid religious option. Certain creationists state outright that the first book of Genesis is unique, as is the God of the Bible, and go so far as to label all other religions "evolutionary" or perverted forms of Biblical revelation. This inability to come to terms with religious pluralism is

Table 1
Illustrative parallel quotations from the General Edition (© 1979; 3rd edition, 3rd printing 1981); and the Public School Edition (© 1978) of *Evolution? The Fossils Say No!* (San Diego: Creation-Life Publishers).

General Edition[a]	Public School Edition[b]
The proponents of this [catastrophist-recent creation] model for interpreting geologic history believe that *the correct interpretation of Genesis* requires acceptance of a creation spanning six 24-hour days. Furthermore, the *genealogies listed in Genesis and elsewhere in the Bible*, it is believed, would restrict the time of creation to somewhere between six thousand and about ten thousand years ago. [italics added]	The proponents of this [catastrophist-recent creation] model for interpreting geological history believe that creation spanned six 24-hour days. Furthermore, it is believed creation occurred thousands rather than billions of years ago.

General Edition[c]	Public School Edition[d]
After many years of intense study of the problem of origins *from both a Biblical and a scientific viewpoint*, I am convinced that the facts of science declare special creation to be the only logical explanation of origins. *'In the beginning God created'. . . . is still the most up-to-date statement that can be made about our origins!* [some italics added]	After many years of intense study of the problem of origins *from a scientific viewpoint*, I am convinced that the facts of science declare special creation to be the only logical explanation of origins. [italics added]

[a]page 60. [b]page 57. [c]pages 186–187. [d]page 174.

behind the effort to teach Biblical creation as science in the public schools. Children from a variety of religious traditions, each with its own idea of origins, must, the creationists believe, learn about "creation," along with its concomitant ideas about materialism, piety, and the "God of the Bible." The creationists' value judgment about the superiority of Christian monotheism and the Genesis creation account is turned into an "alternative" to evolutionary theory, which is identified with every evil in human society.

Creationists are driven by an evangelical and soteriological purpose: they want people to come to salvation in Jesus and must safeguard the literal accuracy of Genesis lest people doubt the story of salvation as well. If the integrity of any part of the Bible is threatened, they believe, then the entire Christian message is threatened. That central concern is evident in the creation-scientists' literature and demonstrates the explicitly religious purpose of their attempt to prove Genesis "scientifically."

The missionary goals of creation-scientists and their legislative advocates are, as a result of the Arkansas litigation, now well-documented. The model creation-science legislation is prepared and disseminated by Paul Ellwanger, founder and head of Citizens for Fairness in Education and a strong Biblical literalist. In private correspondence subpoenaed for use in the Arkansas trial, Ellwanger stated his conviction that teaching evolution in the public schools is the equivalent of using tax dollars against God, but assured his followers that the legislation will "help expose and neutralize such gross errors as Humanism, for without evolution it and all God-less belief systems would simply crumble."[9] The Institute for Creation Research insists that research, writing, and teaching in scientific creationism "is the most effective way in which recognition of God as a sovereign Creator and Savior can be restored" among young people.[10] Luther Sunderland, a clever tactician of the movement, in a confidential memorandum describing how to introduce creation-science into schools by avoiding discussion of religion in school-board presentations, told creationist activists that "presenting the scientific evidences on origins is one of the most effective ways to convince people there is a God, and it can be done without even mentioning the subject."[11] Even Wendell Bird, the creationist lawyer leading the defense of the legislation, explained in a moment of candor why creationists should request instruction in "scientific creationism":

Christians are commanded to be lights for a crooked and perverse nation, and are to stand against the devil with the armor of God (Philippians 2:15; Ephesians 6:11). Christians have a responsibility to ensure light and to oppose evil in the public school system, because our country is shaped powerfully by public school curricula and our tax dollars finance public education.[12]

Question 3: Support Within Secular Scholarship

The fact that course content may be consistent with, or even reinforce, religious beliefs is not sufficient to render its presentation unconstitutional if an independent secular justification exists for its presence in the curriculum. The task of a court is to distinguish between real and pretextual "secular" justifications, and in doing so it must assess the quality of the contribution to non-sectarian education that the "science" offers.

The court should make this assessment by determining the theory's standing within the scientific community as a whole, not by conducting its own inquiry into the theory's merits, a judgment the court may be poorly equipped to make. The court can gauge a theory's standing through examination of the treatment accorded it in recognized journals and by established professional societies and, most importantly, by the extent to which the theory has found its way into the classroom as part of the ordinary process of curriculum development.

The process governing curriculum development is normally in the hands of professional educators, who apply non-religious standards in determining the educational value of proposed course content. These educators look to the scientific community for its assessments of various theories. When this process is left free to operate without political interference, the curriculum accurately mirrors the standing of theories within secular scholarship. Enactment of a state statute or school-board resolution in order to place a theory into the curriculum may strongly indicate that the theory lacks substantial support within the scientific community and that an independent secular justification for teaching the theory does not exist.

Examination of the standing of creation-science in the relevant disciplines provides convincing evidence that it has no substantial support within secular scholarship. Creation-scientists, for example, operate largely outside the scientific community. They publish their claims in books and captive journals disseminated by creationist and other religious publishing houses. With few exceptions, they have not brought their empirical claims to the scientific community for review and criticism in standard journals subject to peer review. When the legislation forced the scientific community to recognize and evaluate the creationists' claims, most major scientific professional societies adopted resolutions rejecting creation-science. And nowhere has creation-science been added to public school curricula because of the sheer force of its ideas. Its inclusion in science classes has always been the result of irregular political intervention in the curriculum development process.

Conclusion

Formal education plays a significant role in shaping a person's world view. In consequence, the courts have been vigilant in assuring that tax-supported education takes place in a religiously neutral environment, and thus efforts to give public schools a sectarian cast are forced to become increasingly sophisticated. Central to most recent efforts has been a decreased emphasis on overt religious references, a description of doctrine as philosophical or scientific, and a careful articulation of an overriding secular objective that purportedly motivates the statute or practice. Such attempts rely upon a sociological phenomenon, the growing influence of science in our society. The American public respects science greatly, even if it does not always understand exactly what science is or how it operates. Any claim that an idea is "scientific" appears to enhance its persuasive power and its public appeal.

"Creation-science" is neither the first nor the last "science" to seek admission to the public schools as a way of encouraging a particular religious perspective. Attempts to screen religiously-inspired advocacy out of tax-supported instruction by developing and applying a legal definition of science are, it seems to me, doomed to failure, at least in the long term. The most serious indictment of a definitional approach is not that philosophers or the courts cannot agree on a definition of science but that any definition will be an insufficiently sensitive caliper for distinguishing propaganda from serious scholarship. No body of information put together solely to persuade students of the superiority of particular religious values, even if it omits explicit religious references and bases its arguments on science, can be taught in the public schools without doing violence to the principles of religious neutrality. A science-based apologetic is nevertheless an apologetic and a functional analysis is more likely to identify that fact.

References

1. See Duane Gish, "Letter to the Editor," 2 *Discover* 6 (July 1981). See also the deposition testimony of Harold G. Coffin, Ariel Roth, Cecil Gerald Van Dyke, and Hilton Fay Hinderliter, quoted in Plaintiffs' Pre-Trial Brief in *McLean v. Arkansas*.

2. *McLean v. Arkansas Board of Education*, 529 F.Supp. 1255 (E.D. Ark. 1982).

3. See, for example, Larry Laudan's commentary in this volume.

4. Laudan, *op. cit.*

5. The description of the religious writings of creation-scientists contained in this article is based on the research and analysis of Linda C. McLain, a para-professional legal assistant with Skadden, Arps, Slate, Meagher & Flom.

6. *1981–1982 General Catalogue*, Christian Heritage College, p. 10.

7. Henry Morris, "Letter to the Editor," 11 *Creation Research Society Quarterly* (December 1974): p. 173.

8. Henry Morris, *Scientific Creationism* (San Diego: Creation-Life Publishers, 1974) (General Edition, 9th Printing, September 1981), pp. 250–251.

9. Private correspondence of Paul Ellwanger, subpoenaed by plaintiffs for use in *McLean v. Arkansas*, Exhibit 6 to Ellwanger deposition, at p. 2.

10. Henry Morris, Letter to new subscribers of *Acts and Facts* (undated) (on file with author).

11. Luther Sunderland, "Introducing Two-Model Teaching of Origins in Public Schools—An Approach That Works," (unpublished, October 1980) (Exhibit 18 to Ellwanger deposition in *McLean v. Arkansas*, p. 8).

12. Wendell Bird, "Evolution in Public Schools and Creation in Students' Homes: What Creationists Can Do," *Impact* Article 70, (April 1979); reprinted in *The Decade of Creation* (San Diego: Creation-Life Publishers, 1981), p. 126, Henry Morris and Donald Rohrer, eds.

The Creation-Science Case and Pro Bono Publico

Peggy L. Kerr

One of the most dramatic and highly publicized moments in the trial of *McLean v. Arkansas Board of Education*[1] occurred when the defendants' expert witness on religion, Norman Geisler, admitted on cross-examination to his belief that UFOs are a "satanic manifestation in the world for the purpose of deception."[2] The cross-examiner who elicited this testimony—and other answers tending to illustrate the extreme and fundamentally unscientific views to which the creationists' literal reading of the Bible can lead—was Anthony J. Siano. Siano, whom press reports generally referred to as "ACLU Attorney Anthony J. Siano," is also an associate at the New York law firm, Skadden, Arps, Slate, Meagher & Flom. He, together with more than a dozen fellow attorneys, numerous law students and still more paralegals, all employed by Skadden, Arps, volunteered his services to the American Civil Liberties Union (ACLU) for the "creation-science" case.[3]

A firm of over 270 attorneys which is best known for its corporate "takeover" work, Skadden, Arps in fact has a broad-based and diversified practice, including general corporate and securities work; commercial and securities litigation; and antitrust, labor, tax, products liability, energy, environmental, bankruptcy, real estate, trusts and estates, employee benefits, equal employment, and election law matters. Most Skadden, Arps litigators hone their skills on "takeover" cases, where the stakes are high, the time pressures severe, and the demands great. Given the substantial involvement of the firm's personnel in the creation-science case, therefore, it was not surprising to read articles asserting that the Arkansas trial was conducted "as if it were a corporate merger."[4]

From May 1981 through December 1981, thousands of hours of the time of Skadden, Arps attorneys, paralegals, and law students were devoted to *McLean v. Arkansas*. Many of those hours could have been spent on fee-generating work; others on getting adequate sleep, reading novels, going to ballgames, or playing with one's children.

Why were the firm and the individuals who volunteered willing to make such a commitment? In part, the answer lies in the ethical tradition of the legal profession, which deems it an obligation to do pro bono publico work. And, as founding partner Leslie H. Arps affirms, "it is also a tradition of this firm. We have *always* encouraged public service work."

Developing the Case

Out of respect for the diversity of political, social, and religious views held by its attorneys, Skadden, Arps neither appears as a firm in any pro bono matter nor lends its institutional imprimatur to one side or another in such cases. The resources of the firm are made available to support pro bono activities, but each individual attorney involved makes his or her own decision on whether to undertake a particular matter (subject to clearance by the firm to prevent conflicts of interest). Thus, the Skadden, Arps attorneys who worked on the creation-science case did so as volunteers, because they wanted to.

The creation-science case first came to the attention of the firm through associate Gary E. Crawford. As a law student at Vanderbilt University, Crawford had worked on a successful legal challenge to a Tennessee statute that had required, in essence, all theories of origins to be taught in public schools and all to be labeled "theories" except the Genesis account of creation.[5] After graduation, Crawford continued his interest in church-state relations, serving on the national advisory board of the Americans United for Separation of Church and State. Upon learning of the enactment of Arkansas Act 590 and of the American Civil Liberties Union's proposed challenge to that statute, Crawford contacted the ACLU to volunteer his services. The ACLU welcomed Crawford's participation, as well as that of other Skadden, Arps personnel. When such pro bono opportunities present themselves, the customary practice is to circulate a memorandum throughout the firm inviting volunteers. The memo about the creation-science case received a resounding response, identifying a large pool of lawyers interested in the issues and willing to work on the case.

The contribution of these volunteers unquestionably made a substantial difference in the presentation of the plaintiffs' case, for at least two reasons. First, apart from the tradition of pro bono service, the efforts of the Skadden, Arps volunteers reflected a wide range of motives affected by background and collected interests. Second, Skadden, Arps personnel also were able to muster resources that are not generally possessed by individual or small-firm practitioners and might otherwise have been unavailable to the ACLU.

Some of the team members are former teachers and parents of schoolchildren, people who have a personal interest in fighting a law that subverts education. During meetings, one parent/attorney angrily asserted, "You can't teach children lies!," referring in part to the creation-science doctrine that the earth is but 6,000 to 10,000 years old, notwithstanding scientific evidence that it is approximately 4.5 billion years old. This point was echoed by physicist Harold Morowitz, who testified as an expert witness on behalf of the plaintiffs. When asked "Why can't we teach the flat earth theory in the schools?" Morowitz replied drily, "Because it's wrong."[6]

Our numbers included many people who came from religious traditions they hold in profound respect. These lawyers shared, with the clergymen and religious organizations who were plaintiffs in *McLean v. Arkansas*, a deep concern that the injection of religion into public school education will degrade religion itself.[7] (Sandra Kurjiaka, director of the Arkansas ACLU affiliate, illustrated Act 590's threat to religion by describing a woman who called her on a radio talk show and said, "I'm a Christian, and I don't want any public school teachers teaching my children how to be a Christian! I'm the only one who's going to do that!") Two members of the team, a recent law school graduate who was working for Skadden, Arps during the summer, and a paralegal, each had religion-related graduate degrees—an educational background that rarely finds expression in the practice of corporate law.

Also among us were a large number of people with education, background, and experience in technical and scientific areas. It was surprising to learn how many attorneys working on antitrust or securities law or on trusts and estates had received their pre-law training in science and technology. One paralegal on the case was a graduate student in the philosophy of science. The opportunity presented by the creation-science case brought together in a single task these dual interests in science and law.

These talents and enthusiasms were, in addition, well supported by the resources and standard practices of the firm. Early in the case, Skadden, Arps mobilized its forces into teams, each headed and supervised by an experienced attorney. The "religion" team engaged consultants and identified potential expert witnesses and issued subpoenas to various creationist organizations for publications, correspondence, and other pertinent documents. The "science" team also engaged consultants and identified possible expert witnesses in various scientific fields. In addition, utilizing the scientific and technical backgrounds of the attorneys involved, as well as the advice of volunteer science consultants, the members of the science team analyzed the creationists' arguments and generated memoranda setting forth both the arguments and the scientific reasons why they were unfounded and unpersuasive. The "education" team reviewed science-teaching materials now in use which implicate the origin of life and of the universe and those materials the creationists touted as appropriate for use in the teaching of origins. That team, too, engaged consultants and potential expert witnesses, and interviewed and prepared Arkansas teachers and educators for possible testimony at trial.

The "law" team—its numbers swelled by law students who worked for Skadden, Arps during Summer 1981 as "summer associates" and during the school year as "research assistants"—researched every aspect of the law applicable to the case. They authored extensive legal memoranda dealing with the scope of the Establishment Clause, the contours of the right of academic freedom, the meaning of the "vagueness" doctrine of the Fourteenth Amendment, and a variety of other legal issues pertinent to pre-trial discovery and to evidentiary issues expected to arise at trial. These memoranda ultimately formed the basis for the briefs filed with the court.

Fueling the work of each of these teams of lawyers and law students was the factual research performed largely by paralegals. Paralegals, of whom Skadden, Arps employs more than a hundred, are highly qualified college graduates, often with special training in legal research and procedures; some are committed to a career in paralegal work, others intend to work only for a few years while they are making decisions about graduate or professional schools, for example. In the Arkansas case, the paralegals collected what is perhaps the most complete creationist library outside of the creationist organizations themselves, identifying and ordering books and periodicals, getting themselves on mailing lists, even attending (by invitation of the organizations involved) creationist conferences. They launched a "reading project" in which the major creationist works were analyzed for religious themes and religious references. Where appropriate (for example, to collect information about persons that the Arkansas Attorney General had identified as potential expert witnesses for defendants), they fanned out to public and science libraries and utilized the firm's computer research facilities, including its access to many data bases, to gather background information. They found, for example, that some creationists believe in UFOs, a fact eventually used to dramatic advantage in the cross-examination of Geisler described above.

When the time came for pre-trial discovery, the attorneys were well-armed. Under the rules of procedure which apply to cases in Federal courts, parties are usually entitled to take "deposition" testimony (i.e., testimony under oath but outside the presence of the judge) and thus to avoid surprises at the trial itself. One Skadden, Arps research assistant was deployed full-time to find scientists, preferably located near the site of the deposition, whose areas of specialization were akin to those of the witnesses and who would attend the deposition to provide on-the-spot analysis of the witnesses' claims and to feed questions to the attorney taking the testimony. As a result of that assistant's labors— and the overwhelming support of the scientific community—approximately one hundred scientists advised (also on a pro bono basis) the litigation team, and one or more experts in the witness's own field attended each deposition of a creation-science witness.

The discovery schedule agreed to by plaintiffs and defendants required Skadden, Arps attorneys to be traveling for over two weeks, taking approximately twenty depositions in four cities. Only a few days into the schedule, the Arkansas Attorney General unilaterally suspended discovery. Within a day's time, by having paralegals locate and reserve hotel rooms as deposition sites, and by calling upon colleagues at other law firms in different cities, Skadden, Arps arranged to have eleven subpoenas issued and served in ten cities by the following day—a process that was terminated only when the Attorney General agreed to resume depositions on a voluntary basis.

As "subpoena day" illustrates, law firms such as Skadden, Arps are poised for quick action, to an extent probably greater than at places less accustomed to the strenuous demands of "takeover" litigation. The attorneys and employees are schooled in work habits that rarely respect the 9-to-5 workday. When a client's needs require weekend work, or when passable preparation can be accomplished by 5 p.m. but thorough preparation will take until midnight, the lawyer involved will be in the office (or in the hotel room) tying up the last loose ends. This cast of mind, applied no less to the creation-science case than to a multimillion-dollar tender offer, had a payoff at trial. As reported in the *Washington Post*:

It is true that the ACLU and the New York firm of Skadden, Arps attacked the Arkansas law with a powerful case. Their brief is so good that there is talk of publishing it as a book. Their witnesses gave brilliant little summaries of several fields of science, history and religious philosophy.

They cross-examined the witnesses for the other side—impassioned believers, rebellious educators and scientific oddities—with devastating results.[8]

Apparent to anyone who has tried to translate science, history, or religious philosophy into legalese (or back again) is that neither a publishable brief nor the witnesses' "brilliant little summaries" could have been accomplished without days, weeks, or months of dedicated work by many people with libraries, computers, word processors, and portable airline guides at their command.*

As another motivation for all this effort, one ought not to discount the allure of a case that raises major constitutional questions, in which a settlement was unlikely and a trial on the merits virtually assured. Trial experience is a precious commodity in a large firm, and experience trying a case that the world views as "historic" is eagerly sought. But cutting across the individual motives was the fact that *McLean v. Arkansas* was simply a fascinating case, one that implicated the most deeply treasured of all individual freedoms: the right to worship and to believe according to one's own conscience, and the right to study and to learn according to one's own capacities. Although plaintiffs and their counsel battled a law that might have resulted in an entire generation of scientific illiterates, they also rallied to uphold the First Amendment's guarantee that no one will have another's religion forced upon him by the government, either openly or in disguise.

Pro Bono Publico

The legal profession is currently engaged in some controversy over the scope and extent of a lawyer's obligation to do pro bono publico work and the appropriate definition of such work. Few professionals disagree

*As Bruce J. Ennis, Jr., former Legal Director of the ACLU and now in the firm of Ennis, Friedman, Bersoff & Ewing, Washington, DC, put it, "The lawyers who worked on the case were not just lawyers; they were *excellent* lawyers." — PK

that, besides serving their own pocketbooks, attorneys must also serve the public weal. In the nineteenth century, this obligation was expressed in a magniloquently phrased resolution:

To my clients I will be faithful; and in their causes zealous and industrious. Those who can afford to compensate me, must do so; but I shall never close my ear or my heart because my client's means are low. Those who have none, and who have just causes, are, of all others, the best entitled to sue, or be defended; and they shall receive a due portion of my services, cheerfully given.[9]

The intraprofessional controversy over the appropriate contours of the pro bono obligation arises in the context of the American Bar Association's current thoroughgoing revision of the Code of Professional Responsibility.[10] Even those who argue most vigorously for a mandatory requirement that each individual lawyer devote a minimum number of hours per year to pro bono activities have acknowledged the social utility of "collective" efforts. In such cases as the Skadden, Arps' involvement in the creation-science case, some among a group of attorneys commit far more than the minimum numbers of hours, with the support of the group as a whole.[11] Organizations such as the ACLU, the NAACP Legal Defense Fund, or the New York Lawyers for the Public Interest, which frequently generate complex legal matters, depend heavily on collective efforts of law firms.[12] Thus, the argument in favor of providing incentives for collective efforts is that such a practice

would facilitate the generation of useful legal services by causing firms to encourage more extensive pro bono activities by those individuals who are most motivated to work in the area. It would also facilitate the handling of complex matters requiring the work of a group of lawyers, by encouraging firms to undertake such matters as a firm project. Supervision of younger lawyers by partners or other experienced lawyers would be encouraged, and administration would be assisted. . . . In other respects, it would accentuate the treatment of public service matters as firm obligations in the same manner as fee-generating cases are handled.[13]

Even in the absence of formal incentives for collective pro bono endeavors, lawyers in firms such as Skadden, Arps operate in a climate where such efforts in fact are undertaken, are met with favor by the employers, and are highly valued by the organizations and clients to whom the legal services are donated. Every young associate at Skadden, Arps (as at similar firms) sooner or later hears the stories that set the tone for pro bono work, acquired as part of the "oral history" of a firm. For example, Leslie Arps spent years as counsel to the New York

State Crime Commission where, among other inquiries, he supervised the waterfront investigation memorialized in Elia Kazan's *On the Waterfront*. One hears, too, of Joseph H. Flom's early battles against organized anti-Semitism and his representation, with the ACLU, of the "Communist Ten" during the McCarthy era. Labor lawyer John D. Feerick (soon to leave this firm to become Dean of the Fordham University School of Law) has done extensive work in the areas of presidential succession and electoral college reform and is credited with having lent his "expert guiding hand throughout the process of crafting the (Twenty-Fifth) Amendment, obtaining its approval by Congress and ratification by the states, and implementing it for the first time."[14] The tales also include how Litigation Practice Leader Barry H. Garfinkel—a director of the Legal Aid Society—spent nine years of nights and weekends litigating a major environmental case.[15] And there are innumerable other stories of individuals who hold public office, take leaves of absence to work in the public sector, represent indigent defendants or charitable, educational, and cultural organizations, or hold directorships of such organizations.

In a sense, the more senior lawyers are the trendsetters or role models. In the creation-science case, they were also, in many instances, a valuable cheering squad, eager for news of the progress of the case, reading with advice, counsel, and encouragement—always bearing the message that devotion of enormous effort to an issue of such public significance was an activity very much approved of.

Despite its public visibility the creation-science case was only a piece of a much larger pro bono commitment by Skadden, Arps personnel, and pro bono activities in the entire legal community.[16] In early 1981, for example, a number of the firm's attorneys and paralegals, working with the ad hoc New York Lawyers' Committee to Preserve Legal Services, aided in the preparation of a brief for Congress, urging the reauthorization and continued funding of the Legal Services Corporation. That brief—an inordinately impressive document—contains the essence of the spirit of pro bono commitment. "To destroy the (Legal Services Corporation) now," that brief argued,

would be a signal that the nation will demand respect for law from all, but will deliver its protections only to those who can pay. No such signal should be sent by the government of a people whose pledge of allegiance, recited daily by millions, boasts of "liberty and justice for all."

Skadden, Arps pioneered a program under which a firm associate works full-time for the New York City Law Department for a period of six months, while continuing to receive a salary from the firm. In addition, Skadden, Arps attorneys regularly undertake the representation of complainants in employment-discrimination cases, having been deputized by the New York State Division of Human Rights; and, for many years, other colleagues have participated actively in the Volunteer Division of the Legal Aid Society (formerly the Community Law Offices), representing the poor in a variety of matters ranging from domestic relations to criminal proceedings. They also volunteer to work on legal matters under the auspices of the New York Lawyers for the Public Interest. These endeavors, plus the firm's arrangement of the Second Circuit Judicial Conference, service by individuals on committees of the Association of the Bar of the City of New York, and a variety of other matters, make for a large and active pro bono program representative of the commitment made by the profession as a whole.

In accepting pro bono service as a professional obligation, the legal profession seems to assume as an underlying premise that the obligation can be discharged in a way that is satisfying, rather than noisome. The voluntary involvement of so many attorneys, law students, and paralegals in the creation-science case lends suport to that proposition — representing, as it does, the happy confluence of a legal dispute of public importance and a sizable team of qualified professionals willing, and indeed pleased, to tackle it.

Notes and References

1. *McLean v. Arkansas Board of Education*, 529 F.Supp. 1255 (E.D. Ark. 1982).

2. The quotation is approximate.

3. Although the focus of this article is on Skadden, Arps' involvement in the creation-science case, the representation of the plaintiffs in this case was a joint effort, with major roles played by Bruce J. Ennis, Jr., and Jack D. Novik of the national office of the American Civil Liberties Union and by Robert M. Cearley, Jr., and Philip E. Kaplan of the Arkansas Bar.

4. Gene Lyons, "Repealing the Enlightenment," *Harper's Magazine* (April 1982): 84. The context of the quotation is Lyons' criticism of the plaintiffs in *McLean v. Arkansas* for having allowed themselves to be "muzzled by a platoon of lawyers," rather than matching the Moral Majority and the Institute for Creation Research "in volume and stridency" of public statements about the trial.

5. *Steele v. Waters*, 527 S.W.2d 72 (Tenn. 1975).

6. The quotation is approximate.

7. See *Engel v. Vitale*, 370 U.S. 421, 431 (1962) (The "first and most immediate purpose [of the Establishment Clause of the First Amendment] rested on the belief that a union of government and religion tends to destroy government and to degrade religion").

8. Philip Hilts, "Creation Trial: Less Circus, More Law," *Washington Post* (21 December 1981): A3.

9. David Hoffman, *A Course of Legal Study*, Resolution XVIII (2nd ed., 1936), reprinted in H. Drinker, *Legal Ethics* (1953). In the early nineteenth century, Hoffman, of the Baltimore Bar, wrote a series of "resolutions" for the "assistance of the young practitioner."

10. In August 1975, the American Bar Association House of Delegates approved the recommendation of the ABA Special Committee on Public Interest Practice, resolving "that it is a basic professional responsibility of each lawyer engaged in the practice of law to provide public interest legal services" without fee, or at a substantially reduced fee, in one or more of the following areas: poverty law, civil rights law, public rights law, charitable organization representation, and administration of justice.

In 1980, the ABA Commission on Evaluation of Professional Standards proposed a mandatory requirement for unpaid public interest service, coupled with an annual reporting requirement. ABA Commission on Evaluation of Professional Standards, *Model Rules of Professional Conduct*—Discussion Draft (30 January 1980), Rule 9.1, p. 118.

That proposal was withdrawn, and the later Working Draft provides:
A lawyer should render public interest legal service. A lawyer may discharge this responsibility by providing professional service at no fee or a reduced fee to persons of limited means or to public service or charitable groups or organizations, or by service in activities for improving the law, the legal system or the legal profession.
ABA Commission on Evaluation of Professional Standards, *Model Rules of Professional Conduct*—Working Draft (21 March 1981), Rule 6.1, p. 224.

11. Special Committee on the Lawyer's Pro Bono Obligations of the Association of the Bar of the City of New York, *Toward a Mandatory Contribution of Public Service Practice by Every Lawyer* (1979), p. 19.

12. *Ibid.*, p. 52, note 17.

13. *Ibid.*, p. 19.

14. Birch Bayh, "Forward," J. Feerick, *The Twenty-Fifth Amendment* (1976), p. ix.

15. See *Greene County Planning Board v. Federal Power Commission*, 455 F.2d 412 (2d Circuit), *cert.* denied 409 U.S. 849 (1972).

16. Lawyers from other large New York law firms participate, or have participated, in many of the programs described herein. Lawyers from the firm Paul, Weiss, Rifkind, Wharton & Garrison, for example, have been working with the ACLU in a challenge to a Lousiana statute that is substantially similar to Arkansas Act 590.

Creationism, Censorship, and Academic Freedom

Susan P. Sturm

Fighting Creationism in the public schools is essential to the preservation of First Amendment rights guaranteed by the U.S. Constitution. The real issue raised by the campaign to teach Creationism in the public schools is religious freedom. In *McLean v. Arkansas Board of Education* and other cases throughout the country, the courts have shown that the covert aim of this campaign is to use creationist doctrine to mask the introduction of religious teaching, contrary to the First-Amendment prohibition against any "law respecting an establishment of religion." As Judge Overton noted in *McLean v. Arkansas*, "The Establishment Clause enshrines two central values: voluntarism and pluralism. And it is in the area of the public schools that these values must be guarded most vigilantly." Many years ago, Justice Frankfurter had captured the particular importance of the Establishment Clause in the context of the public schools:

Designed to serve as perhaps the most powerful agency for promoting cohesion among a heterogeneous democratic people, the public school must keep scrupulously free from entanglement in the strife of sects. The preservation of the community from divisive conflicts, of Government from irreconcilable pressures by religious groups, of religion from censorship and coercion however subtly exercised, requires strict confinement of the State to instruction other than religious, leaving to the individual's church and home, indoctrination in the faith of his choice. [*McCollum v. Board of Education*, 333 U.S. 203, 216–217 (1948)]

"Creation-science" is rooted in a profound desire to circumvent the First Amendment. The creationists' own writings, in fact, attest to their campaign to insinuate religious doctrines in the public schools. For example, a funding letter from the Creation Science Research Center contains the following language: "We already have a State mandated religion of atheism—of Godlessness—of satanism—and no church training of one hour a week will overcome this onslaught of anti-God teaching in the classroom. The church must get involved." Such clear rejection of the concept of neutrality embodied in the Establishment Clause is evident throughout the creation-science literature.

The creationists, conscious that this position is not legally or politically defensible, have attempted instead to frame the Creationism debate in terms more palatable to the general public. Judge Overton found this to be the case in Arkansas. The proponents of Creationism showed an awareness that Act 590—the Creationism Act—"is a religious crusade, coupled with a desire to conceal this fact." By developing a pseudo-science that plays on the public's fear and ignorance of science, the creationists thus disguise their true purpose. They avoid direct references to God, use labels and catch-phrases to disguise the religious aspects, and publish differing public and private-school editions of their materials.

How, then, do the creationists approach lawyers, scientists, and educators who are unlikely to be misled by the pseudo-science? First, the creationists have attempted (and to some extent successfully) to confound the legal and academic communities by posing the creation-science issue in terms of censorship and academic freedom. This approach ignores and distorts the meaning of those concepts, and is but a thinly veiled challenge to the First Amendment.

When attention is focused not on the political statements and characterizations of creationists, but instead on the substance of creation-science, it is apparent that there is no merit to the creationists' charge of censorship. The claim is, in fact, based on the faulty and dangerous assumption that creation-science is science and not religion. This assumption cannot withstand the slightest scrutiny.

Nowhere was this clearer than in the courtroom in Arkansas. There, a group of parents, teachers, and clergy brought an action challenging the constitutionality of Act 590, which required "balanced treatment" of "creation-science" and "evolution-science." In the courtroom, the proponents of Act 590 could not rely on labels, catch-phrases, or political sympathies. Their claims and contentions were subject to scrutiny under the rules of evidence and the legal standards developed to determine violations of the Establishment Clause of the First Amendment:

First, the statute must have a secular legislative purpose; second, its principal or primary effect must be one that neither advances nor inhibits religion; finally, the statute must not foster "an excessive government entanglement with religion."

Following this standard, plaintiffs undertook to prove that creation-science is religion, and that there is no secular justification for including it in the public school curriculum.

Enactment of Act 590 was religiously motivated; creation-science is not science, and it has no educational merit as science. But it was the creationists themselves who provided proof of the religious purpose, either through their correspondence (introduced as evidence) or through the testimony of the proponents of Act 590. Eminent theologians testified about the nature of Fundamentalism and the religious character of creation-science. Reputable scientists testified concerning creation-science's failure to meet the criteria of science. Prominent educators testified concerning the harmful effect of teaching creation-science.

The state, seeking to uphold the constitutionality of Act 590, called a series of witnesses who identified themselves as creation scientists and claimed to show that creation-science is genuinely scientific. Most of their testimony consisted of criticism of various scientific theories they labeled as "evolution-science." Based on their critique of molecular genetics, paleontology, radiometric dating, astronomy, and other scientific disciplines, the witnesses concluded that evolution was disproven, and that this in turn proves the validity of the creation theory embodied in the Book of Genesis. Indeed, most of the creation-scientists who testified had, as a condition for membership in a creation-science association, taken an oath stating that the Bible is the written word of God, and its assertions historically and scientifically true.

A theologian testifying for the state attempted to minimize the religious content of creation-science by drawing a distinction between the belief in God and the belief that God exists, a distinction which, as the Court found, is contrary to common understanding and contradicts settled case law.

The judge's opinion clearly demonstrates that creation-science is religion, not science. He found that the evidence "establishes that the definition of 'creation-science' contained in [the Act] has as its unmentioned reference the first eleven chapters of the Book of Genesis." He further found that "the two-model approach of the creationists is simply a contrived dualism that has no scientific factual basis or legitimate educational purpose." He concluded that creation-science has no scientific merit or educational value as science, and that "the only real effect of Act 590 is the advancement of religion."

It is in this light that the creationists' contention that keeping Creationism out of the public schools violates academic freedom and constitutes censorship—a contention that has no basis in law or fact—must be evaluated. The defendants in the Arkansas case attempted to suggest that the scientific community censors creation scientists. The judge put this assertion to rest, noting that no witness produced evidence of a scientific article that had been submitted to a scientific journal and been rejected. Thus, the creationists did not sustain the claim of censorship.

More importantly, claims of academic freedom and censorship are red herrings. To the extent that creationists' work presents valid and scientific critiques, hypotheses, or experiments, it may be included in the science curriculum; but the critique of science—to the extent that there is any validity to it—is simply the first step of creation-science. The second step is the leap to Genesis, which embodies the organizing principle of Creationism.

The First Amendment exists to prevent indoctrination of a particular religious view. By prohibiting the teaching of Creationism in the public schools, the courts are not implying that science is more correct or more important than religion or any other way of understanding the world. Nor does the First-Amendment prohibition stifle creationist thought or publication of their ideas. They have substantial resources and intellectual freedom. "Creationism" may be (and, indeed, is) vigorously presented and advocated in the appropriate forum. Rather, the exclusion of creation-science from the public schools preserves the rights of individuals to be free from the imposition of a state-sponsored orthodoxy. Preservation of First-Amendment principles to which the American Civil Liberties Union is dedicated thus requires opposition to the teaching of creation-science in the public schools.

The Creationist Controversy: The Interrelation of Inquiry and Belief

Langdon Gilkey

In calling for "balanced treatment" in science classes of "evolution-science" and "creation-science," the Arkansas "Creationism" law (Act 590 of 1981) tried to identify very broadly each of these "scientific models." Although the law did not mention either God or the Book of Genesis, it did list as the elements of the creation-science model: the sudden creation of all things out of nothing; creation of separate and distinct "kinds" at the beginning; the catastrophic interpretation of the earth's history; and the "recent" beginning of the universe. Each of these elements requires for its intelligibility the notion of a supranatural creative act. The definition of creation-science also asserted "the insufficiency of natural selection and mutations" to explain development, and "the naturalistic [read "atheistic"] character" of evolution-science in the emergence of life forms. Act 590 declared its purpose as that of insuring "free debate" and "neutrality" in the teaching of origins by preventing the "dogmatic" teaching of *one* model (evolution) and consequent establishment of theological liberalism and/or atheistic humanism. The subject of origins is in fact taught, the Act pointed out, but only with some evolutionary/naturalistic interpretation. Because this latter model is no self-evident "fact of science," the Act argued, and if academic freedom for teachers and students alike is to be preserved, an "alternative scientific model" should be presented.

Confusions Over Scientific Inquiry

The basic error reflected in the Arkansas law is to regard these two models—one religious (Creationism) and the other scientific (evolution)—as equivalent, logically comparable, and mutually exclusive theories or interpretations. Creationist documents present both as parallel explanations of "origins." Paradoxically, both are argued to be "scientific" or "equally scientific," yet both are called equally "religious"—one model representing a believing, Christian and Biblical religion and the other espousing an "atheistic" or "humanistic" religion. The person who believes in evolution cannot also believe in God or the Bible, this attitude argues; correspondingly, a person who believes in God, and finds some sort of truth in Genesis, must deny evolution. Only as equivalent hypotheses can both models be regarded as scientific explanations, can one model be regarded as the representative of valid religion, and can the State be viewed as having an apparently legitimate responsibility to accommodate both. This error of regarding evolution and Genesis as comparable, mutually exclusive "explanations" is not confined to the fundamentalist community; it represents the confusion out of which this controversy as a whole has arisen. The following two points descriptive of the creationists' arguments contain innumerable further confusions about not only religion but also science, especially its formal structure, methods, and canons.

First, the creationists claim that Creationism is not religion but science—a "scientific model" based on "scientific evidence" or "scientific facts" and thus "at least as scientific as evolution." It is not religion because it neither appeals to scriptural or doctrinal authority nor talks about God "religiously" (e.g., as a personal, loving savior and so on). Neither creation nor evolution, they argue, can be "scientifically proved" because origins, however interpreted, lie beyond direct observation and so beyond experimental testing. Thus, the relative status of each model depends on the "scientific evidence" to which each can appeal, and the capacity of each model to "explain intelligibly" these data.

Second, creationists' writings, and the Arkansas law itself, assume that what the creationists call a "naturalistic explanation"—and, by extension, any theory or science based on such an explanation—is inherently atheistic. If a theory leaves out God—as the scientific theory of evolution certainly does—then it is clearly false and so cannot *really* be "science."* Evolution, therefore, to them means not merely a tentative theory full of empirical or scientific "holes"; it represents a de-

---

*In a scientific culture, science is regarded not only as *true* but also as *defining truth*. As a consequence, some religious people in such a culture can believe that an atheistic science must *ipso facto* be not science but *false science* (i.e., they do *not* say that science is false but that *this* is false science.—LG

liberate and powerful expression of naturalism or atheism. And in some of the creationists' religious writings, the same authors portray evolution as an instrument of aggression by cosmic forces of evil, by the Devil himself.

What sorts of confusions are represented here? First, there is confusion about what science is. The creationists—many of whom are trained scientists—speak of "scientific facts" and "scientific evidence"; they see science as located in its facts, rather than in its theoretical structure. There is little recognition of the canons of scientific method, the logical conditions that make a theory a part of science; and creation-science contravenes each of these major canons. The creation-science "model" is, therefore, not an example of science at all: it involves a supra-natural cause, transcendent to the system of finite causes; it explains in terms of purposes and intentions; and it cites a transcendent, unique, and unrepeatable—even in principle, uncontrollable—action. It represents, therefore, logically and linguistically, a re-edition of a familiar form— that is, natural theology, which argues that certain data point "rationally" to a philosophical/religious conclusion, namely, to the agency of a divine being.

Second, the creationists fail to distinguish the question of *ultimate* origins (Where did it *all* come from?) from the quite different question of *proximate* origins (How did A arise out of B, if it did?). They ignore the (scholastic) distinction between the *primary* causality of a First Cause, with which philosophy or theology might deal, and *secondary* causality, which is causality confined to finite factors. Assuming that it is science's role to deal with the truth and, therefore, with *all* of the truth, they conclude that a scientific explanation of origins must be an *exhaustive* explanation and must be inclusive of all possible related factors or causes. If evolution theory deals with proximate origins, it must also deal with the question of ultimate origins. If, in this process, evolution theory has left out God, then it must be asserting that there is no God, or that the divine is in no way the Creator of the process of secondary causes. At the Arkansas trial, the creationists therefore interpreted the scientific witnesses' demurrals that "science does not raise the question about God at all" as meaning that science rules out the presence of God in any way.

The creationists ignore—or possibly are unaware of—the restrictive canons of the scientific method (e.g., that no super-natural causes may be included in a theory) and the distinction of ultimate from proximate origins. As a consequence, they fail (as do many) to understand that, although science provides testable and relatively certain conclusions, its conclusions or answers are limited and not exhaustive. As in the parallel cases of historical inquiry, law, or psychology, the "atheism" of natural science is a priori and methodological. No acceptable historical or legal hypothesis can include the divine as a central cause of a historical event or of a crime. If, as another example, objective psychological inquiry cannot locate experimentally (behavioristically) any sign of my inward freedom, this does not mean either that such freedom is not, in fact, there or that psychological inquiry denies human intentionality— unless, of course, one specifies in advance that the results of *only* such inquiry are to be considered "real." In the same way, scientific explanations of proximate origins are confined to using *finite* causes as principles of explanation and thus leave quite open the question of God. The charge that evolution is "atheistic" is, as a consequence, a simple tautology, an analytic judgment equivalent to the assertion "this is a scientific theory."

Confusions about the nature and rules of the scientific method, about the distinction between scientific and other forms of knowing, and so about the limitations of scientific inquiry—and the subsequent distinction of ultimate from proximate questions of origins—have bred the theoretical confusions that made this case possible. These same confusions are not confined to fundamentalist groups. In a culture in which science represents the paradigmatic, if not the exclusive, mode of knowing, knowledge is apt to be regarded as all on one level. As a consequence, scientific and religious explanations are bound to conflict and may be regarded at one moment as "science" and at another as "religion." At the same time that Naturalistic Humanism parades itself as "what we scientists now know," therefore, fundamentalist Creationism challenges evolution science with the claim to represent an *alternative* "science."

The Theological Testimony

The logic of the plaintiffs' case in *McLean v. Arkansas* was directed at the confusions embodied in the Act 590 itself and in the "scientific" defense of the act. The plaintiffs wished, first of all, to establish that the creationist model represents a particular form of religion, one exclusive of not only all non-Christian religions but also most recognized forms of Christian faith. The theological testimony at the trial dealt with these arguments *theoretically*.

In a monotheistic culture, the testimony went, all that is religious has to do with God and all that has to do with God is religious. For monotheistic religions, God is the principle of ultimate reality and therefore is the source of all other reality, the principle of authority in revelation, the source of every religious way of life, and the founding agent of the religious community. Religion refers essentially and exclusively to God. This religious reference includes *all* of God's actions, His or Her creative activity in establishing the world, and divine redeeming action in reuniting with us. For this reason, all supra-natural beings who create are as much "gods" as are all such beings who save. To speak of a creator of all things, therefore, is to speak religiously, even if a philosophical argument may also be produced to give secular warrants for this notion. Not all religions have gods and surely not all worship God, but all that has to do with God is certainly religious.

When, in order to circumvent this argument, the authors of Act 590 sought to separate the Creator implied by creation-science from the notion of "religion," it was ironic that they came close thereby to the "first and worst" Christian heresy—the denial of monotheism (i.e., the belief in two gods, one of them the morally dubious creator and the other the good, loving Savior God). It was no accident, as I noted in my testimony, that the first article of the earliest Christian creed witnessed to "one God, the Father Almighty, and Maker of Heaven and earth." Further, if religious statements are referent to God and not to finite causes—and this defines most Western religious assertions—then, of all the statements about God that could be made, the proposition that God creates "out of nothing" is the *most* religious. By definition, no other agent was present because this act of creation established all other agents. Act 590, therefore, is religious not because it refers explicitly to a doctrine or appeals to scripture, but because the notion of the agency of a supra-natural being is essential to each of the constitutive elements of the creationist model—and that is, ipso facto, religious speech. Finally, the creationist model proposes a *particular* religious view of creation, different from that of other religious traditions and different from other Christian interpretations. It is not a scientific model at all but a theological one and thus (as Judge Overton held in his opinion) contravenes the First Amendment.

Testimony by scientists at the Arkansas trial supported this line of arguement. Creation-science is not, they said, science at all. In making its own case as legitimate science, Creationism has misunderstood the methods of science, many of its fundamental laws, and many of its present theoretical conclusions. In denying that evolution is valid science, in asserting that it is disguised religion, and in rejecting the testability of evolution theory's major conclusions, the creation scientists reveal that they do not understand how the relevant sciences proceed, how theories are composed and tested, and what scientific status various hypotheses can claim. The distinction between alternative *scientific* accounts and a *religious* account of origins was fundamental to the plaintiffs' strategy. Time and again, scientists testified "but *that* is not science"—a statement dependent for its force on the mostly unexplicated distinction between scientific and religious speech and understanding, a distinction operative in types of questions asked, in procedures and authorities invoked, in the forms of speech used, and in the shape of the resulting system of symbols.

At the Interface of Inquiry and Belief

The Arkansas case does *not* represent simple warfare between the enlightened forces of science on the one hand and the darkening forces of religion on the other—as fundamentalists, the secular intelligentsia, and the media have frequently implied. For example, the plaintiffs in the Arkansas case included almost all the main-line churches and only *one* scientific organization; half of the plaintiffs' witnesses represented religion. Ironically, perhaps, scientists provided the central testimony for the defense.

Historians of science now recognize that the image of a warfare between science and religion was not even true in the nineteenth century. Technology and science today characterize *all* levels of society and, correspondingly, so do "faith" and religion. Moreover, the religion represented in this case is itself a product of that technological and scientific society. Fundamentalist and cultic forms of religion have grown in our lifetime *because of*, not in spite of, the dilemmas of a technological society. Both technology and religion are permanent and essential aspects of the culture as a whole; both are potentially creative and potentially destructive.

If this litigation is not merely the last episode in a continuing contest but is, in fact, a much more complex problem of misinterpretation, on many levels, of two essential and pervasive aspects of cultural life, then how are we to understand it? First, much of the blame for misinterpretation rests with the churches and the schools of theology. One of theology's major tasks in the last two centuries has been to understand reflectively how religious faith (and, by extension, Christian religious faith—even in Genesis) can be reinterpreted in the light of modern science. Yet, a satisfactory (i.e., intelligible) understanding of the relationship between religion and science has not permeated American church life (or, I might add, all of American society). Many people still assume that to believe in God or the Bible one must reject the notion of evolution. For example, following my testimony in Arkansas, a *Time* reporter asked me, "If you are a Christian theologian and believe in revelation, how can you accept Darwin?" I replied, "On many counts I don't. I understand there are today a number of scientific reasons for questioning elements of Darwin's theory." The magazine reporter seemed even more baffled by my reply.

The responsibility of the religious community is clear enough. What is not so obvious is the scientific community's responsibility for this same problem. Many scientists share with the fundamentalists the confused notion that so-called religious knowledge and scientific knowledge exist on the same level and that, as science advances, scientific knowledge simply replaces and dissolves religious myth. Religion is viewed primarily as "belief," an early and very shaky stage of human cognition and understanding, and as an enterprise that reaches its culmination in modern science. Religion is thus regarded as "pre-science," "early science," or "primitive science," and can be expected to vanish, as do all denizens of the night, when the daylight of science appears and spreads—a view theologians often characterize as the "Walt Disney theory" of cultural history. While scientists who believe this interpretation reject *all* of religion as pre-science, the creationists cling to certain aspects of science associated with their religious faith and reject only scientific hypotheses that compete with the doctrines of that faith. One encounters this view of science as dissolving religious truth in the writings of Julian Huxley, Gaylord Simpson, Jacob Bronowski, and Carl Sagan; a recent volume of readings in evolution theory edited by C. Leon Harris, for example, classifies Genesis under the heading of "prescientific myth" and cites the great St. Augustine under the bizarre title "The Infanticide of Science: Augustine and the Dark Ages." This is about as informed and sensitive as listing Einstein and Fermi under the heading "The Development of Destructive Weaponry."

Not only do the scientific naturalists and the fundamentalists agree in theory that religious truth and scientific theory are direct competitors and so mutually exclusive, but each perspective tends to breed and encourage the other. Much of Scientific Naturalism has gestated out of parental Fundamentalism or orthodoxy. Correspondingly, the new fundamentalist reaction against evolution has arisen in part because of the frequently careless and uninformed way evolution-science is being taught. Each time a child comes home and reports, "I learned in school today that Genesis is wrong," the seed is planted for creationist reaction. As Fundamentalism originally arose in the late nineteenth and early twentieth centuries as a reaction against liberal and modernist Protestantism, so creation-science has arisen in our day in reaction to Scientific Naturalism, which is a global (and, therefore, "religious") Weltanschauung based on science but extending beyond science to encompass and frame all of human experience.

In a scientific culture it is rightly taken for granted that the modern educated person should understand at least enough science to be aware of its methods and limits, of its most general conclusions, and, above all, of the "world" of reality, truth, and value which science implies. Because of the myth of the absoluteness and self-sufficiency of scientific knowing and because of a confidence in the imminent disappearance of religion, the reverse has not been the case. Scientists have not been expected to be aware of the wider implications of their methods and, by and large, have not been encouraged to reflect on the relations of scientific truth to *other* ways of knowing such as historical inquiry, arts, morals, philosophy, or religion.

The myth that religion will vanish in a secular and scientific culture is itself vanishing under the pressure of repeated historical falsification. Part of the counterfactual data is the reappearance of Fundamentalism, the appearance (in unexpected variety and strength) of non-Western religious cults, and the appearance in political and economic ideologies of historical myths that unify, empower, and direct modern technological societies much as traditional religions unified and directed archaic societies—for example, the myth of democratic, liberal, and scientific Progress or the myth of Marxist Communism. None of these new forces represents traditional religion carried over from the past, as do Western Christianity and Judaism. These new religious forms appear and reappear *out of* and *because of* a scientific, technological culture in response, first, to the demand for a credible system of symbols giving structure, meaning, and direction to nature, history, society, and the self, and, second, to the particular sharp anxieties—and even terrors—of a technological age, especially one in apparent, although not admitted, decline.

The creative as opposed to the destructive effects of scientific knowledge and technology *depend* on other aspects of culture: on its political and legal structures and processes, on its moral integrity and courage, and on the forms of its religious faith. Our century has also shown the persistence, the permanence, the ever-renewed power, as well as the deep ambiguity of religion. But religion in one form or another does and will continue to exist—like science—in both demonic *and* creative form. The relations between these two essential and permanent elements of culture represent a recurrent, fundamental issue that should be a part of the training and self-understanding of both the scientific and the religious communities. In such times as these, the religious dimension tends to expand, and, unfortunately, to grow in fanaticism, intolerance, and violence; science and technology likewise tend to concentrate more and more on developing greater and greater means of destructive and repressive power. The combination represents a most dependable recipe for self-destruction. Let each of us rethink the role of our own profession in this light and especially its relation to other communities in our total social life.

The Scientific Response to   Joel Cracraft
Creationism

Introduction

The rising public awareness of Creationism during the past several years derives not only from the creationists' highly efficient public-relations program, but also from a vigorous response by scientists, philosophers, and theologians. No new tactics were needed to persuade conservative evangelicals that Creationism deserved a wider hearing within secular society, particularly in the public school system; the tenets of Creationism have always been acceptable to these and similar fundamentalist religious groups. But the creationists have also had considerable success in persuading a substantial "uncommitted" portion of Americans—many as unaware of contemporary thinking on science, philosophy, and theology as of the content of Creationism—that Creationism possesses sufficient merit to warrant public support, however marginal. The creationists have accomplished this success primarily through well-placed political pressure, but their major argument has been that Creationism is at least as scientific as the present-day scientific disciplines and thus should be treated with fairness, an open mind, and a willingness to consider equally all "alternative" scientific ideas.

For educated persons without formal training in the sciences, many creationist statements—cloaked by the jargon of science and pronounced with conviction by people with Ph.D.'s from reputable universities—have an authoritative ring. Especially when such statements are made in the forum of public debate, the opposing scientist, rarely as skilled in forensics as the creationist, often appears to be on the defensive. The success of the campaign to persuade the public that Creationism has legitimate scientific content has had nothing to do with the normal process of evaluating the merits of competing scientific ideas. The creationists ask the public to arbitrate what is claimed to be a scientific debate while scientists seek judgment of their ideas primarily within the scientific community.

In this essay I will examine the proposition that Creationism is legitimate science. To do that, I must first consider, at least briefly, the philosophical basis of that claim, including the methods the creationists use to discredit contemporary scientific thought. Most of my essay, however, will be directed at evaluating the six "scientific" claims most often asserted by the creationists in their writings and public debates.

The Philosophical Position of Creationism

Because they accept the Bible as a true, factual account of the history of the universe, earth, and life, the creationists are compelled logically to redefine the philosophical and methodological foundations of science. No longer is science characterized solely by naturalistic explanations of natural phenomena—the creationists place supernatural explanations within the legitimate domain of science. No longer can science construct explanatory hypotheses about events having a time dimension—to creationists, science must study only the observable, only that which can be verified in a laboratory experiment. No longer must scientific ideas, or conjectures, be subject to criticism and eventual rejection—some statements, such as those derived from revelation, are not only to be considered scientific in their content, but also impervious to criticism. In public forums, the creationists are very reluctant to admit that the aim of creation-science is to fit observations about nature into a literal Biblical framework. But in their writings, the creationists (including such leaders as Henry M. Morris, Director of the Institute for Creation Research) are often quite candid about their motives:

The Bible is the Word of God, absolutely inerrant and verbally inspired . . . The Bible gives us the revelation we need, and it will be found that all the known facts of science or history can be very satisfactorily understood within this Biblical framework.[1]

It is impossible, therefore, to evaluate the "science" in Creationism as if this were a typical scientific debate. The creationist world view dictates suspension of commonly accepted canons of science. Moreover, in their criticisms of contemporary science, the creationists rarely attempt to support their claims on merit alone. Rather, they promote a form of dualism: evidence *against* a contemporary scientific idea is evidence *for* a creationist interpretation. Of course, finding evidence against any important scientific idea is not difficult. Real science—particularly "important" science—is full of debate. By choosing the side they wish to discredit and then selectively quoting opposing scientists, the creationists attempt to construct their case. Consider two among many examples.

In *Creation: The Facts of Life*, Gary Parker of the Institute for Creation Research emphasizes the design apparent in life as evidence of a supernatural designer. He even cites Harvard biologist Richard C. Lewontin to that effect:

Then there's the "marvelous fit of organisms to the environment," the special adaptations of cleaner fish, woodpeckers, bombardier beetles, etc., etc.,—what Darwin called "Difficulties with the Theory," and what Harvard's Lewontin (1978) called "the chief evidence of a Supreme Designer." Because of their "perfection of structure," he says, organisms "appear to have been carefully and artfully designed."[2]

What Parker fails to tell his readers is that Lewontin was, in truth, describing the cultural milieu that Darwin faced as he struggled with the problem of adaptation. Here are Lewontin's words in his *Scientific American* article:

It was the marvelous fit of organisms to the environment, much more than the great diversity of forms, that was the chief evidence of a Supreme Designer. Darwin realized that if a naturalistic theory of evolution was to be successful, it would have to explain the apparent perfection of organisms and not simply their variations.[3]

Yet another example involves Thomas G. Barnes, member of the Institute for Creation Research. Writing in 1973, he claims that the earth's magnetic field does not reverse its polarity (with the North Pole becoming the South, and vice versa).[4] By 1973, of course, geologists and geophysicists had demonstrated conclusively the actuality of reversals. To support his case, Barnes cited one of the leaders in the field, J. A. Jacobs, as expressing doubts about the validity of the reversal hypothesis.[5] The Jacobs' book on which Barnes relied, however, was published in 1963, several years *before* crucial experiments finally demonstrated the reversals, and thus it is no wonder that he was cautious at that time.[6] The Parker and Barnes methods of argumentation are typical of those in all creationist writings.

The Scientific Case Against Creationism

Every scientist who becomes engaged in the creation-evolution controversy experiences a sense of frustration over the constant necessity to defend well-established scientific knowledge against the creationists' claims to the contrary. The creationist challenge has not been formulated from a rigorous debate within the scientific literature. This challenge is strictly external to the scientific community and arises within organizations whose stated purposes are avowedly religious.

In this section I will briefly discuss six of the major arguments that the creationists raise against the findings of modern science. It is impossible within this brief essay to provide references to the vast technical literature concerning these issues. Instead, I will attempt to explain both the nature of the creationists' assertions and the scientific evidence against them.

The Second Law of Thermodynamics
Perhaps no scientific claim is so important to the creationists as their assertion that the laws of physics, especially as embodied in thermodynamics, are more in accord with Creationism than with contemporary physics, chemistry, or biology. They state this repeatedly:

The Second Law of Thermodynamics is especially significant in its support of the creation model and, correspondingly, its contradiction of the evolution model . . . Evolution requires a universal principle of upward change; the entropy law is a universal principle of downward change.[7]

We are warranted, then, in concluding that the evolutionary process . . . is completely precluded by the Second Law of Thermodynamics. There seems no way of modifying the basic evolutionary model to accommodate this Second Law.[8]

The creationists argue that the inherent tendency of thermodynamic systems to pass from states of high order (low entropy) to those less ordered is opposite to and necessarily precludes a long, continuous history of increasing diversification and complexity of life forms, as interpreted by evolutionary biology. The controversy is easily illustrated by an example. Consider a closed container with a mixture of two different gases subjected to a thermal gradient. At the start, let molecules of one kind be in the right half, of the other kind in the other. This would represent a highly ordered configuration for these molecules. If the system is kept closed to outside energy flow, the gases will soon mix completely and the entire system will reach a configuration of least order, or maximum entropy.

When the creationists state that evolution is impossible because it violates the second law, they accept the premise that the system comprising biological materials is a *closed* system. One of the first things taught in elementary biology courses is that the biosphere is, in fact, an open system, receiving a continuous flow of energy from the outside in the form of sunlight. The biosphere, a small region of the universe, may manifest high degrees of order at the expense of the surroundings, ultimately connected to the huge and continuing increase of the entropy of our sun. When this point is brought to the attention of the creationists, they simply respond that it makes no difference. Even an open system, they claim, must have a "program" to direct the increase in order and a "converter" to change the energy needed to build that order. These arguments are, however, straw men manufactured by the creationists. Changes in open systems are predicated upon the configurations already present, and the work applied to systems to drive the change arises inherently from the conversion of one form of energy to another.

During the last thirty or forty years a whole branch of thermodynamics has arisen which treats the irreversible processes within nonequilibrium systems, such as the biosphere.[9] Few scientists today believe that we can use the second law to imply, as do the creationists, that regions of open systems cannot become more organized through time. A closed thermodynamic system is an idealized concept, its closest approximation being a highly controlled experimental situation. Numerous recent studies have successfully applied thermodynamics, properly understood, directly to the question of biological evolution.[10] Not a single creationist has ever challenged this research within the scientific literature.

The Improbability of the Evolution of Life

The second most important argument raised by the creationists actually is closely tied to the question of irreversible thermodynamics. The "laws of probability," the creationists assert, rule out the evolution of life. They pose the question this way:

Assume a "sea" of freely available components, each uniquely capable of performing a specific useful function. What is the probability that two or more of them can come together by chance to form an integrated functioning organism?[11]

The real problem for evolutionists is explaining how a cell in all of its complexity could arise suddenly from simple inorganic atoms . . . The sudden "poof!" formation of a cell would demand a supernatural act by an agent with supernatural power and intelligence.[12]

An extensive body of scientific research exists on the origin of large macromolecules and the cellular origin of life.[13] The creationists have never appeared to be cognizant of this literature, at least on an intellectual and scientific level; instead, they have chosen to re-define the scientific question in a manner that best suits their own polemics. They refer consistently to the origin of complex organic molecules, or to life itself, by characterizing this evolution as being "sudden," "random," and entirely due to "chance." They go to great lengths to calculate the "improbability" that simple molecules could have united spontaneously to form life.

The creationists' probability argument is a caricature of science, not a careful consideration of the complicated problems surrounding this issue. The common elements comprising organic molecules here on earth—hydrogen, oxygen, nitrogen, and carbon—are also characteristic of all bodies of the universe. One of these elements' basic properties is their strong affinity for one another to form compounds of varying complexity. Thousands of scientific papers have addressed the physics and chemistry of these chemical reactions. Many scientists have investigated the kinetics of these reactions—how larger and larger molecules are synthesized naturally—and attempted to relate these reactions to the conditions postulated to have been present on the primitive earth 4 to 4.5 billion years ago. No scientist claims that life arose "spontaneously" from non-life, nor does any scientist profess to having certain knowledge of the precise sequence of chemical reactions leading to macromolecules, such as nucleic acids, characteristic of life.

The creationists' argument is a flagrant misapplication of probability analysis. To determine the combined probability for a series of events, one must know the probability of each event in the series. The creationists arbitrarily designate the probabilities of a series of chemical events, with no regard to scientific relevance. Moreover, they assume that the changes from one state to another occur at random, independent of preceding events, even though it is well-known in the scientific literature that these chemical sequences do not take place randomly. Viewed within the framework of an open thermodynamic system, the second law prescribes an inherent increase in the complexity of the chemical reactions leading to life. A large literature exists on the problem of "self-organization" in molecular evolution, all of which the creationists apparently ignore.

The Age of the Earth

The creationists make two substantive claims about the age of the earth and the methods used to measure it. First, they declare that the geologists' universally accepted dating techniques, in particular radiometric dating, are invalid. Second, they assert that the age of the earth is very much younger (on the order of 10,000 to 20,000 years) than the age supposed by modern science.

Geologists have established, virtually beyond scientific doubt, that the earth is approximately 4.5 to 4.6 billion years old. That the stratigraphic record of sediments can be sequentially dated by radiometric decay rates is not now a matter of question among geologists who study dating techniques. Each radioactive element decays at a unique and constant rate and these rates are not influenced by external factors such as extremes of temperature or pressure. The creationists simply assert that these rates are not constant. Yet, at the Arkansas creation trial, every one of the creationists' geological witnesses—including Robert Gentry, their chief expert on radioactivity—testified that no scientific evidence exists which questions the constancy of these decay rates.

The creationists sometimes invoke a "singularity" at about 6,000 years corresponding to their suggested time for the Noachian "Flood"; at this singularity, the decay rates slowed down significantly—more or less to their present level. Prior to that time, the rates were much higher, thus giving the appearance that the earth is billions of years old, when actually it is only thousands. James Hopson of the University of Chicago has suggested to me a simple response to this supposition of a supernatural event: if the creationists are correct in believing that the earth is only thousands of years old, and that the decay rates at one time accelerated, then the amount of heat released from that amount of radioactive decay would have been sufficient—by a large margin—to have vaporized the earth.

By their own admission, the creationists cannot provide any scientific evidence for such dramatic changes in radioactive decay rates or their assertion that these rates do not measure the true age of the earth. The creationists turn to a supernatural "singularity," which is a belief derived from religion, not from the evidence of science.

The Geological Record
For over two hundred years, geologists have investigated the history of the earth. Many of these workers in the eighteenth and early nineteenth centuries realized that the rate of sedimentary processes observable in nature was too slow to account for the thickness of the earth's strata within the time specified in the Bible. Moreover, they recognized that strata could be correlated with one another over great

distances. Surely, they reasoned, the earth is much older than previously believed.

Out of this inquiry arose different schools of thought: The "Neptunists" interpreted the geological data in terms of a universal ocean or flood, whereas the "Vulcanists" believed volcanic action was instrumental in shaping the earth's surface. But many workers, notably Charles Lyell, saw geological processes operating less catastrophically, and the hypothesis of a very old earth could not be denied much longer. Notably, this transition in thinking about the age of the earth was accomplished by scientists with deeply-held religious convictions. Nevertheless, many of them were willing to be led by empirical science rather than by recourse to supernatural explanation.

In contrast, the modern creationists wish to return to a time when Scripture was considered the final arbiter of knowledge about earth history. The creationist positions on geological issues are so firmly imbedded in religion that it is difficult to address them in rational, scientific terms. Consider the following statement from *The Genesis Flood*, easily the most important creationist publication on geology:

We believe that the Bible . . . gives us the true framework of historical and scientific interpretation . . . This framework is one of special creation of all things, complete and perfect in the beginning, followed by the introduction of a universal principle of decay and death into the world after man's sin, culminating in a worldwide cataclysmic destruction . . . by the Genesis Flood. We take this revealed framework of history as our basic datum, and then try to see how all the pertinent data can be understood in this context.[14]

There is no accepted body of scientific evidence supporting the idea that a universal flood covered all the earth about 6,000 years ago. Recent advances in geology centered around plate tectonics and continental drift, in fact, create special dilemmas for the creationists because of the mass of empirical data supporting a new interpretation of geological history. In dealing with such data, the creationists merely demur, saying that the results *are* compatible with the "Flood," which they see as providing the energy for continental drift.[15] Such a statement is representative of religious apologetics, not rational scientific inquiry.

Fossil Transitions and Patterns of Descent

The strongest argument for the "fact" of evolution—the highly corroborated notion that there exists a historical record of descent with modification—is the non-random stratigraphic sequences of fossil organisms. The creationists make two main attacks on evolutionary interpretation of this record: first, they deny ipso facto that there are fossil transitions between the different kinds of organisms; second, they argue that the stratigraphic sequences of fossil forms are more easily explained by a "Flood" geology than by the accumulation of these remains over a long period of time. Both arguments are demonstrably false.

If you do not want to believe in something, the easiest solution is to deny its existence, and, when that fails, to define it such that it cannot exist. The creationists have done just that with the idea of transitional forms:

Transitional series *must* have existed in the past, if evolution is true, and the fossil record should reveal at least some of these . . . The fact is, however, that no such transitional series—or even occasional transitional forms—have ever been found in the fossil record.[16]

Because there are no transitions between major groups, the creationists conclude, gaps exist just as predicted by the creation model.[17] In making this statement, however, the creationists confuse the idea of "transitional" form and obscure the scientific issue by defining "transitional form" in a way unacceptable to any knowledgeable evolutionist. The creationists argue that a transitional form is one in which all characters are intermediate between two groups. Evolutionists know that characters do not transform simultaneously, but evolve instead at very uneven rates. This variability in rate produces organisms who possess some characteristics (primitive ones) similar to those of their ancestors and others (derived) shared with closely related forms, including possibly their descendants. Each species, then, is an intermediate in some sense of the word; all species possess primitive and derived characters.

The fossil record provides abundant documentation of transitional forms. Many of these fossils yield information about characters shared with earlier forms and those shared with species found later in the fossil record. A comparison among these similarities allows the detection of a pattern of phylogenetic relationships which closely approximates the stratigraphic sequences of fossils. For example, the famous fossil bird *Archaeopteryx* shares many primitive skeletal characters with small bipedal dinosaurs; but it also has some features, such as feathers and a fused "wishbone," that are characteristic of birds. *Archaeopteryx* supplies scientists with a wealth of data about the evolutionary connections between birds and a particular group of reptiles—the dinosaurs. Most evolutionists would not claim, of course, that these intermediate forms are necessarily the direct ancestors of a later group.

The creationists' second assertion about the fossil record is that it accords well with an interpretation based on the "Flood." Creationists point out that, after the "Flood" waters receded, the simpler ocean-dwelling forms would have been buried first, followed by fish and other organisms living near the surface, and then by the terrestrial organisms.[18] But religious apologetics such as this hardly does justice to the complexities of the fossil record in which successively more primitive forms—determined on the basis of morphology and not on their position in time—are found correspondingly lower in the stratigraphic column. If there had been a flood of the proportions described by the creationists, then it is inconceivable that the fossil organisms would have their present distribution in the geological record. A scenario based on the "Flood" predicts that there would be samples of fossil deposits containing mixtures of primitive and advanced organisms such as dinosaurs and modern mammals. This does not happen.

The Argument of Design

Perhaps none of the creationists' arguments is as beguiling as the argument of design. Does not the presence of design stand as evidence of a designer? In their lectures and writings, the creationists frequently cite William Paley's famous watch-maker example, as it appeared in his *Natural Theology* (1802): "The care with which the parts have been made and the fineness of their adjustment can have only one implication, namely, that the watch must have had a maker who understood its construction and who designed it for the use for which it is fitted." Paley, of course, was arguing for the existence of God. Creationists adopt the same argument in what is ostensibly a scientific debate. Of course, design is evidence for God—if one starts with the premise that only God can produce design. In this case, rather than seek to understand the implications of apparent design from a scientific point of view, the creationists turn the problem into an exercise in religious apologetics.

Darwin's solution to the problem of design in nature was his famous theory of adaptation by natural selection. Most evolutionists still subscribe to that explanation, although a growing number are expressing doubts. Hence, the search for an explanation of how organisms come to have the form they do has once again become the subject of considerable research. That science might not be able to provide an entirely satisfactory answer to this question at this time is not, as creationists imply, evidence against an evolutionary point of view, but merely representative of the fact that science never stops questioning.

Conclusion

I have tried to present a brief summary of the nature of the creationists' "science" and how the scientific community responds to it. Even a cursory glance at the creationists' literature demonstrates their religious motives. For example, this passage from a book by Henry Morris:

If we expect to learn anything more than this about the Creation, then God above can tell us. And He has told us! In the Bible, which is the Word of God, He has told us everything we *need* to know about the Creation and earth's primeval history.[19]

By rejecting the normal goals and methods of science and accepting revelation as a valid way to acquire scientific knowledge, the creationists seek to re-interpret currently accepted scientific knowledge through the eyes of religious Fundamentalism.

References

1. Henry M. Morris, *Evolution and the Modern Christian* (Philadelphia: Presbyterian and Reformed Publishing Co., 1967), p. 55.

2. Gary Parker, *Creation: The Facts of Life* (San Diego: Creation-Life Publishers, 1980), p. 144.

3. Richard C. Lewontin, "Adaptation," 239 *Scientific American* 3 (1978): 212–230.

4. Thomas G. Barnes, "Origin and Destiny of the Earth's Magnetic Field," *Institute for Creation Research Technical Monograph No. 4* (1973): 1–64.

5. J. A. Jacobs, *The Earth's Core and Geomagnetism* (New York: The Macmillan Company, 1963).

6. I am grateful to G. Brent Dalrymple for bringing this example to my attention.

7. Henry M. Morris, *The Scientific Case for Creationism* (San Diego: Creation-Life Publishers, 1977), pp. 13–16.

8. Henry M. Morris, ed., *Scientific Creationism* (San Diego: Creation-Life Publishers, 1974), p. 45.

9. See I. Prigogine, *From Being to Becoming* (San Francisco: W. H. Freeman and Company, 1980).

10. See J. S. Wicken, "The Generation of Complexity in Evolution: A Thermodynamic and Information-Theoretical Discussion," 77 *Journal of Theoretical Biology* (1979): 349–365; J. S. Wicken, "A Thermodynamic Theory of Evolution," 87 *Journal of Theoretical Biology* (1980): 9–23; and D. R. Brooks and E. O. Wiley, "Nonequilibrium Evolution: A Theory of Organismic Change," *Nature* (in press, 1982).

11. Henry M. Morris (1974), *op. cit.*, p. 59.

12. R. L. Wysong, *The Creation-Evolution Controversy* (Midland, MI: Inquiry Press, 1976), pp. 410–411.

13. See, for example, H. Noda, ed., *Origin of Life, Proceedings of the Second ISSOL Meeting* (Tokyo, Japan: Japan Scientific Societies Press, 1978); J. F. Fredrick, ed., "Origins and Evolution of Eukaryotic Intracellular Organelles," 361 *Annals of the New York Academy of Sciences* (1981): 1–512.

14. J. C. Whitcomb, Jr., and H. M. Morris, *The Genesis Flood* (Grand Rapids: Baker Book House, 1961), p. xxvi.

15. Henry M. Morris (1974), *op. cit.*, p. 128.

16. Henry M. Morris, *The Scientific Case for Creation* (San Diego: Creation-Life Publishers, 1977), p. 30.

17. Duane T. Gish, *Evolution? The Fossils Say No!* [Public School Edition] (San Diego: Creation-Life Publishers, 1978), p. 65.

18. Whitcomb and Morris, *op. cit.*, pp. 265–276.

19. Henry M. Morris (1967), *op. cit.*, p. 54.

Creation-Science Is Not Science   Michael Ruse

In December 1981 I appeared as an expert witness for the plaintiffs and the American Civil Liberties Union (ACLU) in their successful challenge of Arkansas Act 590, which demanded that teachers give "balanced treatment" to "creation-science" and evolutionary ideas.[1] My presence occasioned some surprise, for I am an historian and philosopher of science. In this essay, I do not intend to apologize for either my existence or my calling, nor do I intend to relive past victories;[2] rather, I want to explain why a philosopher and historian of science finds the teaching of creation-science in science classrooms offensive.

Obviously, the crux of the issue—the center of the plaintiffs' case—is the status of creation-science. Its advocates claim that it is genuine science and may, therefore, be legitimately and properly taught in the public schools. Its detractors claim that it is not genuine science but a form of religion—dogmatic Biblical literalism by another name. Which is it, and who is to decide?

It is somewhat easier to describe who should participate in decisions on this issue. On the one hand, one naturally appeals to the authority of religious people and theologians. Does creation-science fit the accepted definitions of a religion? (In Arkansas, the ACLU produced theologians who said that indeed it did.) One also appeals to the authority of scientists. Does creation-science fit current definitions of science? (In Arkansas, the ACLU produced scientists who said that indeed it did not.)[3]

Having, as it were, appealed to the practitioners—theologians and scientists—a link still seems to be missing. Someone is needed to talk at a more theoretical level about the nature of science—any science—and then show that creation-science simply does not fit the part. As a philosopher and an historian, it is my job to look at science, and to ask precisely those questions about defining characteristics.

What Is Science?

It is simply not possible to give a neat definition—specifying necessary and sufficient characteristics—which separates all and only those things that have ever been called "science." The concept "science" is not as easily definable as, for example, the concept "triangle." Science is a phenomenon that has developed through the ages—dragging itself apart from religion, philosophy, superstition, and other bodies of human opinion and belief.[4]

What we call "science" today is a reasonably striking and distinctive set of claims, which have a number of characteristic features. As with most things in life, some items fall on the borderline between science and nonscience (e.g., perhaps Freudian psychoanalytic theory). But it is possible to state positively that, for example, physics and chemistry are sciences, and Plato's Theory of Forms and Swedenborgian theology are not.[5]

In looking for defining features, the obvious place to start is with science's most striking aspect—it is an empirical enterprise about the real world of sensation. This is not to say that science refers only to observable entities. Every mature science contains unobservables, like electrons and genes, but ultimately, these unobservables refer to the world around us. Science attempts to understand this empirical world. What is the basis for this understanding? Surveying science and the history of science today, one thing stands out: science involves a search for order. More specifically, science looks for unbroken, blind, natural regularities (*laws*). Things in the world do not happen in just any old way. They follow set paths, and science tries to capture this fact. Bodies of science, therefore, known variously as "theories" or "paradigms" or "sets of models," are collections of laws.[6]

Thus, in Newtonian physics we find Newton's three laws of motion, the law of gravitational attraction, Kepler's laws of planetary motion, and so forth. Similarly, for instance, in population genetics we find the Hardy-Weinberg law. However, when we turn to something like philosophy, we do not find the same appeal to empirical law. Plato's Theory of Forms only indirectly refers to this world. Analogously, religion does not insist on unbroken law. Indeed, religious beliefs frequently allow or suppose events outside law or else events that violate law (miracles). Jesus feeding the 5,000 with the loaves and fishes was one such event. This is not to say that religion is false, but it does say that religion is not science. When the loaves and fishes multiplied to a sufficiency to feed so many people, things happened that did not obey natural law, and hence the feeding of the 5,000 is an event beyond the ken of science.[7]

A major part of the scientific enterprise involves the use of law to effect *explanation*. One tries to show why things are as they are—and how they fall beneath or follow from law (together perhaps with certain specified initial conditions). Why, for example, does a cannon ball go in a parabola and not in a circle? Because of the constraints of Newton's laws. Why do two blue-eyed parents always have blue-eyed children? Because this trait obeys Mendel's first law, given the particular way in which the genes control eye-color. A scientific explanation must appeal to law and must show that what is being explained had to occur. The explanation excludes those things that did not happen.[8]

The other side of explanations is *prediction*. The laws indicate what is going to happen: that the ball will go in a parabola, that the child will be blue-eyed. In science, as well as in futurology, one can also, as it were, predict backwards. Using laws, one infers that a particular, hitherto-unknown phenomenon or event took place in the past. Thus, for instance, one might use the laws of physics to infer back to some eclipse of the sun reported in ancient writings.

Closely connected with the twin notions of explanation and prediction comes *testability*. A genuine scientific theory lays itself open to check against the real world: the scientist can see if the inferences made in explanation and prediction actually obtain in nature. Does the chemical reaction proceed as suspected? In Young's double slit experiment, does one find the bands of light and dark predicted by the wave theory? Do the continents show the expected aftereffects of drift?

Testability is a two-way process. The researcher looks for some positive evidence, for *confirmation*. No one will take seriously a scientific theory that has no empirical support (although obviously a younger theory is liable to be less well-supported than an older theory). Conversely, a theory must be open to possible refutation. If the facts speak against a theory, then it must go. A body of science must be *falsifiable*. For example, Kepler's laws could have been false: if a planet were discovered going in squares, then the laws would have been shown to be incorrect. However, no amount of empirical evidence can disprove, for example, the Kantian philosophical claim that one ought to treat people as ends rather than means. Similarly, Catholic religious claims about transubstantiation (the changing of the bread and wine into the body and blood of Christ) are unfalsifiable.[9]

Science is *tentative*. Ultimately, a scientist must be prepared to reject his theory. Unfortunately, not all scientists are prepared to do in practice what they promise to do in theory; but the weaknesses of individuals are counterbalanced by the fact that, as a group, scientists do give up theories that fail to answer to new or reconsidered evidence. In the last thirty years, for example, geologists have reversed their strong convictions that the continents never move.

Scientists do not, of course, immediately throw their theories away as soon as any counter-evidence arrives. If a theory is powerful and successful, then some problems will be tolerated, but scientists must be prepared to change their minds in the face of the empirical evidence. In this regard, the scientists differ from both the philosophers and the theologians. Nothing in the real world would make the Kantian change his mind, and the Catholic is equally dogmatic, despite any empirical evidence about the stability of bread and wine. Such evidence is simply considered irrelevant.[10]

Some other features of science should also be mentioned, for instance, the urge for simplicity and unification; however, I have now listed the major characteristics. Good science—like good philosophy and good religion—presupposes an attitude that one might describe as professional *integrity*. A scientist should not cheat or falsify data or quote out of context or do any other thing that is intellectually dishonest. Of course, as always, some individuals fail; but science as a whole disapproves of such actions. Indeed, when transgressors are detected, they are usually expelled from the community. Science depends on honesty in the realm of ideas. One may cheat on one's taxes; one may not fiddle the data.[11]

Creation-Science Considered

How does creation-science fit the criteria of science listed in the previous section? By "creation-science" in this context, I refer not just to the definition given in Act 590, but to the whole body of literature which goes by that name. The doctrine includes the claims that the universe is very young (6,000 to 20,000 years), that everything started instantaneously, that human beings had ancestry separate from apes, and that a monstrous flood once engulfed the entire earth.[12]

Laws and Natural Regularities

Science is about unbroken, natural regularity. It does not admit miracles. It is clear, therefore, that again and again, creation-science invokes happenings and causes outside of law. For instance, the only reasonable inference from Act 590 (certainly the inference that was accepted in the Arkansas court) is that for creation-science the origin of the universe and life in it is not bound by law. Whereas the definition of creation-science includes the unqualified phrase "sudden creation of the universe, energy and life from nothing," the definition of evolution specifically includes the qualification that its view of origins is "naturalistic." Because "naturalistic" means "subject to empirical law," the deliberate omission of such a term in the characterization of creation-science means that no laws were involved.

In confirmation of this inference, we can find identical claims in the writings of creation scientists: for instance, the following passage from Duane T. Gish's popular work *Evolution? The Fossils Say No!*

Creation. By creation we mean the bringing into being of the basic kinds of plants and animals by the process of sudden, or fiat, creation described in the first two chapters of Genesis. Here we find the creation by God of the plants and animals, each commanded to reproduce after its own kind using processes which were essentially instantaneous.

We do not know how God created, what processes He used, *for God used processes which are not now operating anywhere in the natural universe.* This is why we refer to divine creation as special creation. We cannot discover by scientific investigations anything about the creative processes used by God.[13]

By Gish's own admission, we are not dealing with science. Similar sentiments can be found in *The Genesis Flood* by John Whitcomb, Jr., and Henry M. Morris:

But during the period of Creation, God was introducing order and organization and energization into the universe in a very high degree, even to life itself! *It is thus quite plain that the processes used by God in creation were utterly different from the processes which now operate in the universe!* The Creation was a unique period, entirely incommensurate with this present world. This is plainly emphasized and reemphasized in the divine revelation which God has given us concerning Creation, which concludes with these words: "And the heavens and the earth were *finished,* and *all* the host of them. And on the seventh day God *finished* His work which He had made; and He *rested* on the seventh day from *all His work* which He had made. And God blessed the seventh day, and hallowed it; because that in it He *rested* from *all* his work which God had created and made." In view of these strong and repeated assertions, is it not the height of presumption for man to attempt to study Creation in terms of present processes?[14]

Creation scientists generally acknowledge *The Genesis Flood* to be the seminal contribution that led to the growth of the creation-science movement. Morris, in particular, is the father figure of creation-science and Gish his chief lieutenant.

Creation scientists also break with law in many other instances. The creationists believe that the Flood, for example, could not have just occurred through blind regularities. As Whitcomb and Morris make very clear, certain supernatural interventions were necessary to bring about the Flood.[15] Similarly, in order to ensure the survival of at least some organisms, God had to busy himself and break through law.

Explanation and Prediction
Given the crucial role that physical laws play for the scientist in these processes, neither explanation nor prediction is possible where no law exists. Thus, explanation and prediction simply cannot even be attempted when one deals with creation-science accounts either of origins or of the Flood.

Even against the broader vistas of biology, creation-science is inadequate. Scientific explanation/prediction must lead to the thing being explained/predicted, showing why that thing obtains and not other things. Why does the ball go in a parabola? Why does it not describe a circle? Take an important and pervasive biological phenomenon, namely, "homologies," the isomorphisms between the bones of different animals. These similarities were recognized as pervasive facets of nature even before Darwin published *The Origin of Species*. Why are the bones in the forelimbs of humans, horses, whales, and birds all so similar, even though the functions are quite different? Evolutionists explain homologies naturally and easily, as a result of common descent. Creationists can give no explanation, and make no predictions. All they can offer is the disingenuous comment that homology signifies nothing, because classification is all man-made and arbitrary anyway. Is it arbitrary that man is not classified with the birds?[16] Why are Darwin's finches distributed in the way that we find on the Galapagos? Why are there fourteen separate species of this little bird, scattered over a small group of islands in the Pacific on the equator? On those rare occasions when Darwin's finches do fly into the pages of creation-science, it is claimed either that they are all the same species (false), or that they are a case of degeneration from one "kind" created back at the beginning of life.[17] Apart from the fact that "kind" is a term of classification to be found only in Genesis, this is no explanation. How could such a division of the finches have occurred, given the short span that the creationists allow since the Creation? And, in any case, Darwin's finches are anything but degenerates. Different species of finch have entirely different sorts of beaks, adapted for different foodstuffs—evolution of the most sophisticated type.[18]

Testability, Confirmation, and Falsifiability

Testability, confirmation, and falsifiability are no better treated by creation-science. A scientific theory must provide more than just after-the-fact explanations of things that one already knows. One must push out into the frontiers of new knowledge, trying to predict new facts, and risking the theory against the discovery of possible falsifying information. One cannot simply work at a secondary level, constantly protecting one's views against threat: forever inventing ad hoc hypotheses to save one's core assumptions.

Creation scientists do little or nothing by way of genuine test. Indeed, the most striking thing about the whole body of creation-science literature is the virtual absence of *any* experimental or observational work by creation scientists. Almost invariably, the creationists work exclusively with the discoveries and claims of evolutionists, twisting the conclusions to their own ends. Argument proceeds by showing evolution (specifically Darwinism) wrong, rather than by showing Creationism right.

However, this way of proceeding—what the creationists refer to as the "two-model approach"—is simply a fallacious form of argument. The views of people like Fred Hoyle and N. C. Wickramasinghe, who believe that life comes from outer space, are neither creationist nor truly evolutionist.[19] Denying evolution in no way proves Creationism. And, even if a more straightforward either/or between evolution and Creationism existed, the perpetually negative approach is just not the way that science proceeds. One must find one's own evidence in favor of one's position, just as physicists, chemists, and biologists do.

Do creation scientists ever actually expose their theories and ideas to test? Even if they do, when new counter-empirical evidence is discovered, creation scientists appear to pull back, refusing to allow their position to be falsified.

Consider, for instance, the classic case of the "missing link"—namely, that between man and his ancestors. The creationists say that there are no plausible bridging organisms whatsoever. Thus, this super-gap between man and all other animals (alive or dead) supposedly underlines the creationists' contention that man and apes have separate ancestry. But what about the australopithecines, organisms that paleontologists have, for most of this century, claimed are plausible, human ancestors? With respect, argue the creationists, australopithecines are not links, because they had ape-like brains, they *walked* like apes, and they used their knuckles for support, just like gorillas. Hence, the gap remains.[20]

However, such a conclusion can be maintained only by blatant disregard of the empirical evidence. *Australopithecus afarensis* was a creature with a brain the size of that of an ape which walked upright.[21] Yet the creationists do not concede defeat. They then argue that the *Australopithecus afarensis* is like an orangutan.[22] In short, nothing apparently makes the creationists change their minds, or allow their views to be tested, lest they be falsified.

Tentativeness

Creation-science is not science because there is absolutely no way in which creationists will budge from their position. Indeed, the leading organization of creation-science, the Creation Research Society (with over five hundred full members, all of whom must have an advanced degree in a scientific/technological area), demands that its members sign a statement affirming that they take the Bible as literally true.[23] Unfortunately, an organization cannot require such a condition of membership, and then claim to be a scientific organization. Science must be open to change, however confident one may feel at present. Fanatical dogmatism is just not acceptable.

Integrity

Creation-scientists use any fallacy in the logic books to achieve their ends. Most particularly, apart from grossly distorting evolutionists' positions, the creation scientists frequently use inappropriate or incomplete quotations. They take the words of some eminent evolutionist, and attempt to make him or her say exactly the opposite to that intended. For instance, in *Creation: The Facts of Life*, author Gary E. Parker constantly refers to "noted Harvard geneticist" Richard Lewontin as claiming that the hand and the eye are the best evidence of God's design.[24] Can this reference really be true? Has the author of *The Genetic Basis of Evolutionary Change*[25] really foresworn Darwin for Moses? In fact, when one looks at Lewontin's writings, one finds that he says that *before Darwin*, people believed the hand and the eye to be the effect of direct design. Today, scientists believe that such features were produced by the natural process of evolution through natural selection; but a reader learns nothing of this from Parker's book.

Conclusion

What are the essential features of science? Does creation-science have any, all, or none of these features? My answer to this is none. By every mark of what constitutes science, creation-science fails. And, although it has not been my direct purpose to show its true nature, it is surely there for all to see. Miracles brought about by an intervening supervising force speak of only one thing. Creation "science" is actually dogmatic religious Fundamentalism. To regard it as otherwise is an insult to the scientist, as well as to the believer who sees creation-science as a blasphemous distortion of God-given reason. I believe that creation-science should not be taught in the public schools because creation-science is not science.

Notes and References

1. In fact, Act 590 demanded that *if* one teaches evolution, *then* one must also teach creation-science. Presumably a teacher could have stayed away from origins entirely—albeit with large gaps in some courses.

2. For a brief personal account of my experience, see Michael Ruse, "A Philosopher at the Monkey Trial," *New Scientist* (1982): 317–319.

3. Judge William Overton's ruling on the constitutionality (or, rather, unconstitutionality) of Act 590 gives a fair and full account of the various claims made by theologians (including historians and sociologists of religion) and scientists.

4. In my book, *The Darwinian Revolution: Science Red in Tooth and Claw* (Chicago: University of Chicago Press, 1979), I look at the way science was breaking apart from religion in the nineteenth century.

5. What follows is drawn from a number of basic books in the philosophy of science, including R. B. Braithwaite, *Scientific Explanation* (Cambridge, England: Cambridge University Press, 1953); Karl R. Popper, *The Logic of Scientific Discovery* (London: Hutchinson, 1959); E. Nagel, *The Structure of Science* (London: Routledge and Kegan Paul, 1961); Thomas S. Kuhn, *The Structure of Scientific Revolutions* (Chicago: University of Chicago Press, 1962); and C. G. Hempel, *The Philosophy of Natural Science* (Englewood Cliffs, NJ: Prentice-Hall, 1966). The discussion is the same as what I provided for the plaintiffs in a number of position papers. It also formed the basis of my testimony in court,

and, as can be seen from Judge Overton's ruling, was accepted by the court virtually verbatim.

6. One sometimes sees a distinction drawn between "theory" and "model." At the level of this discussion, it is not necessary to discuss specific details. I consider various uses of these terms in my book, *Darwinism Defended: A Guide to the Evolution Controversies* (Reading, MA: Addison-Wesley, 1982).

7. For more on science and miracles, especially with respect to evolutionary questions, see my *Darwinian Revolution, op. cit.*

8. The exact relationship between laws and what they explain has been a matter of much debate. Today, I think most would agree that the connection must be fairly tight—the thing being explained should follow. For more on explanation in biology, see Michael Ruse, *The Philosophy of Biology* (London: Hutchinson, 1973); and David L. Hull, *The Philosophy of Biological Science* (Englewood Cliffs, NJ: Prentice-Hall, 1974). A popular thesis is that explanation of laws involves deduction from other laws. A theory is a body of laws bound in this way: a so-called "hypothetico-deductive" system.

9. Falsifiability today has a high profile in the philosophical and scientific literature. Many scientists, especially, agree with Karl Popper, who has argued that falsifiability is *the* criterion demarcating science from non-science (see especially his *Logic of Scientific Discovery*). My position is that falsifiability is an important part, but only one part, of a spectrum of features required to demarcate science from non-science. For more on this point, see my *Is Science Sexist? And Other Problems in the Biomedical Sciences* (Dordrecht, Holland: Reidel, 1981).

10. At the Arkansas trial, in talking of the tentativeness of science, I drew an analogy in testimony between science and the law. In a criminal trial, one tries to establish guilt "beyond a reasonable doubt." If this can be done, then the criminal is convicted. But, if new evidence is ever discovered that might prove the convicted person innocent, cases can always be reopened. In science, too, scientists make decisions less formally but just as strongly—and get on with business, but cases (theories) can be reopened.

11. Of course, the scientist as citizen may run into problems here!

12. The key definitions in Arkansas Act 590, requiring "balanced treatment" in the public schools, are found in Section 4. Section 4(a)(b) does not specify exactly how old the earth is supposed to be, but in court a span of 6,000 to 20,000 years emerged in testimony.

The fullest account of the creation-science position is given in Henry M. Morris, ed., *Scientific Creationism* (San Diego: Creation-Life Publishers, 1974).

13. Duane T. Gish, *Evolution? The Fossils Say No!* (San Diego: Creation-Life Publishers, 1973), pp. 22–25, his italics.

14. John Whitcomb, Jr., and Henry M. Morris, *The Genesis Flood* (Philadelphia: Presbyterian and Reformed Publishing Company, 1961), pp. 223–224, their italics.

15. *Ibid.*, p. 76.

16. See Morris, *op. cit.*, pp. 71–72, and my discussion in *Darwinism Defended, op. cit.*

17. For instance, in John N. Moore and H. S. Slusher, *Biology/A Search for Order in Complexity* (Grand Rapids: Zondervan, 1977).

18. D. Lack, *Darwin's Finches* (Cambridge, England: Cambridge University Press, 1947).

19. Fred Hoyle and N. C. Wickramasinghe, *Evolution from Space* (London: Dent, 1981).

20. Morris, *op. cit.*, p. 173.

21. Donald Johanson and M. Edey, *Lucy: The Beginnings of Humankind* (New York: Simon and Schuster, 1981).

22. Gary E. Parker, *Creation: The Facts of Life* (San Diego: Creation-Life Publishers, 1979), p. 113.

23. For details of these statements, see the Opinion in *McLean v. Arkansas*, footnote 7.

24. Parker, *op. cit.* See, for instance, pp. 55 and 144. The latter passage is worth quoting in full:
Then there's 'the marvelous fit of organisms to the environment,' the special adaptations of cleaner fish, woodpeckers, bombardier beetles, etc., etc.,—what Darwin called 'Difficulties with the Theory,' and what Harvard's Lewontin (1978) called 'the chief evidence of a Supreme Designer.' Because of their 'perfection of structure,' he says, organisms 'appear to have been carefully and artfully designed.'
The pertinent article by Richard Lewontin is "Adaptation," *Scientific American* (September 1978).

25. Richard C. Lewontin, *The Genetic Basis of Evolutionary Change* (New York: Columbia University Press, 1974).

Commentary on Ruse: Larry Laudan
Science at the Bar—Causes
for Concern

In the wake of the decision in the Arkansas Creationism trial, the
friends of science are apt to be relishing the outcome. The creationists
quite clearly made a botch of their case and there can be little doubt
that the Arkansas decision may, at least for a time, blunt legislative
pressure to enact similar laws in other states. Once the dust has settled,
however, the trial in general and Judge William R. Overton's ruling in
particular may come back to haunt us; for, although the verdict itself
is probably to be commended, it was reached for all the wrong reasons
and by a chain of argument which is hopelessly suspect. Indeed, the
ruling rests on a host of misrepresentations of what science is and how
it works.

The heart of Judge Overton's opinion is a formulation of "the essential
characteristics of science." These characteristics serve as touchstones
for contrasting evolutionary theory with Creationism; they lead Judge
Overton ultimately to the claim, specious in its own right, that since
Creationism is not "science," it must be religion. The opinion offers
five essential properties that demarcate scientific knowledge from other
things: "(1) It is guided by natural law; (2) it has to be explanatory by
reference to natural law; (3) it is testable against the empirical world;
(4) its conclusions are tentative, i.e., are not necessarily the final word;
and (5) it is falsifiable."

These fall naturally into two families: properties (1) and (2) have to
do with lawlikeness and explanatory ability; the other three properties
have to do with the fallibility and testability of scientific claims. I shall
deal with the second set of issues first, because it is there that the most
egregious errors of fact and judgment are to be found.

At various key points in the opinion, Creationism is charged with
being untestable, dogmatic (and thus non-tentative), and unfalsifiable.

All three charges are of dubious merit. For instance, to make the inter-linked claims that Creationism is neither falsifiable nor testable is to assert that Creationism makes no empirical assertions whatever. That is surely false. Creationists make a wide range of testable assertions about empirical matters of fact. Thus, as Judge Overton himself grants (apparently without seeing its implications), the creationists say that the earth is of very recent origin (say 6,000 to 20,000 years old); they argue that most of the geological features of the earth's surface are diluvial in character (i.e., products of the postulated worldwide Noachian deluge); they are committed to a large number of factual historical claims with which the Old Testament is replete; they assert the limited variability of species. They are committed to the view that, since animals and man were created at the same time, the human fossil record must be paleontologically co-extensive with the record of lower animals. It is fair to say that no one has shown how to reconcile such claims with the available evidence—evidence which speaks persuasively to a long earth history, among other things.

In brief, these claims are testable, they have been tested, and they have failed those tests. Unfortunately, the logic of the opinion's analysis precludes saying any of the above. By arguing that the tenets of Cre-ationism are neither testable nor falsifiable, Judge Overton (like those scientists who similarly charge Creationism with being untestable) de-prives science of its strongest argument against Creationism. Indeed, if any doctrine in the history of science has ever been falsified, it is the set of claims associated with "creation-science." Asserting that Crea-tionism makes no empirical claims plays directly, if inadvertently, into the hands of the creationists by immunizing their ideology from empirical confrontation. The correct way to combat Creationism is to confute the empirical claims it does make, not to pretend that it makes no such claims at all.

It is true, of course, that some tenets of Creationism are not testable in isolation (e.g., the claim that man emerged by a direct supernatural act of creation). But that scarcely makes Creationism "unscientific." It is now widely acknowledged that many scientific claims are not testable in isolation, but only when embedded in a larger system of statements, some of whose consequences can be submitted to test.

Judge Overton's third worry about Creationism centers on the issue of revisability. Over and over again, he finds Creationism and its ad-vocates "unscientific" because they have "refuse[d] to change it re-gardless of the evidence developed during the course of the[ir] investigation." In point of fact, the charge is mistaken. If the claims of modern-day creationists are compared with those of their nineteenth-century counterparts, significant shifts in orientation and assertion are evident. One of the most visible opponents of Creationism, Stephen Gould, concedes that creationists have modified their views about the amount of variability allowed at the level of species change. Creationists do, in short, change their minds from time to time. Doubtless they would credit these shifts to their efforts to adjust their views to newly emerging evidence, in what they imagine to be a scientifically respectable way.

Perhaps what Judge Overton had in mind was the fact that some of Creationism's core assumptions (e.g., that there was a Noachian flood, that man did not evolve from lower animals, or that God created the world) seem closed off from any serious modification. But historical and sociological researches on science strongly suggest that the scientists of any epoch likewise regard some of their beliefs as so fundamental as not to be open to repudiation or negotiation. Would Newton, for instance, have been tentative about the claim that there were forces in the world? Are quantum mechanicians willing to contemplate giving up the uncertainty relation? Are physicists willing to specify circumstances under which they would give up energy conservation? Numerous historians and philosophers of science (e.g., Kuhn, Mitroff, Feyerabend, Lakatos) have documented the existence of a certain degree of dogmatism about core commitments in scientific research and have argued that such dogmatism plays a constructive role in promoting the aims of science. I am not denying that there may be subtle but important differences between the dogmatism of scientists and that exhibited by many creationists; but one does not even begin to get at those differences by pretending that science is characterized by an uncompromising openmindedness.

Even worse, the ad hominem charge of dogmatism against Creationism egregiously confuses doctrines with the proponents of those doctrines. Since no law mandates that creationists should be invited into the classroom, it is quite irrelevant whether they themselves are close-minded. The Arkansas statute proposed that Creationism be taught, not that creationists should teach it. What counts is the epistemic status of Creationism, not the cognitive idiosyncrasies of the creationists. Because many of the theses of Creationism are testable, the mind set of creationists has no bearing in law or in fact on the merits of Creationism.

What about the other pair of essential characteristics which the *McLean* opinion cites, namely, that science is a matter of natural law and explainable by natural law? I find the formulation in the opinion to be rather fuzzy; but the general idea appears to be that it is inappropriate and unscientific to postulate the existence of any process or fact which cannot be explained in terms of some known scientific laws—for instance, the creationists' assertion that there are outer limits to the change of species "cannot be explained by natural law." Earlier in the opinion, Judge Overton also writes "there is no scientific explanation for these limits which is guided by natural law," and thus concludes that such limits are unscientific. Still later, remarking on the hypothesis of the Noachian flood, he says: "A worldwide flood as an explanation of the world's geology is not the product of natural law, nor can its occurrence be explained by natural law." Quite how Judge Overton knows that a worldwide flood "cannot" be explained by the laws of science is left opaque; and even if we did not know how to reduce a universal flood to the familiar laws of physics, this requirement is an altogether inappropriate standard for ascertaining whether a claim is scientific. For centuries, scientists have recognized a difference between establishing the existence of a phenomenon and explaining that phenomenon in a lawlike way. Our ultimate goal, no doubt, is to do both. But to suggest, as the *McLean* opinion does repeatedly, that an existence claim (e.g., there was a worldwide flood) is unscientific until we have found the laws on which the alleged phenomenon depends is simply outrageous. Galileo and Newton took themselves to have established the existence of gravitational phenomena, long before anyone was able to give a causal or explanatory account of gravitation. Darwin took himself to have established the existence of natural selection almost a half-century before geneticists were able to lay out the laws of heredity on which natural selection depended. If we took the *McLean* opinion criterion seriously, we should have to say that Newton and Darwin were unscientific; and, to take an example from our own time, it would follow that plate tectonics is unscientific because we have not yet identified the laws of physics and chemistry which account for the dynamics of crustal motion.

The real objection to such creationist claims as that of the (relative) invariability of species is not that such invariability has not been explained by scientific laws, but rather that the evidence for invariability is less robust than the evidence for its contrary, variability. But to say as much requires renunciation of the opinion's other charge—to wit, that Creationism is not testable.

I could continue with this tale of woeful fallacies in the Arkansas ruling, but that is hardly necessary. What is worrisome is that the opinion's line of reasoning—which neatly coincides with the predominant tactic among scientists who have entered the public fray on this issue—leaves many loopholes for the creationists to exploit. As numerous authors have shown, the requirements of testability, revisability, and falsifiability are exceedingly *weak* requirements. Leaving aside the fact that (as I pointed out above) it can be argued that Creationism already satisfies these requirements, it would be easy for a creationist to say the following: "I will abandon my views if we find a living specimen of a species intermediate between man and apes." It is, of course, extremely unlikely that such an individual will be discovered. But in that statement the creationist would satisfy, in one fell swoop, all the formal requirements of testability, falsifiability, and revisability. If we set very weak standards for scientific status—and, let there be no mistake, I believe that all of the opinion's last three criteria fall in this category—then it will be quite simple for Creationism to qualify as "scientific."

Rather than taking on the creationists obliquely and in wholesale fashion by suggesting that what they are doing is "unscientific" *tout court* (which is doubly silly because few authors can even agree on what makes an activity scientific), we should confront their claims directly and in piecemeal fashion by asking what evidence and arguments can be marshalled for and against each of them. The core issue is not whether Creationism satisfies some undemanding and highly controversial definitions of what is scientific; the real question is whether the existing evidence provides stronger arguments for evolutionary theory than for Creationism. Once that question is settled, we will know what belongs in the classroom and what does not. Debating the scientific status of Creationism (especially when "science" is construed in such an unfortunate manner) is a red herring that diverts attention away from the issues that should concern us.

Some defenders of the scientific orthodoxy will probably say that my reservations are just nitpicking ones, and that—at least to a first order of approximation—Judge Overton has correctly identified what is fishy about Creationism. The apologists for science, such as the editor of *The Skeptical Inquirer*, have already objected to those who criticize this whitewash of science "on arcane, semantic grounds . . . [drawn]

from the most remote reaches of the academic philosophy of science."[1] But let us be clear about what is at stake. In setting out in the *McLean* opinion to characterize the "essential" nature of science, Judge Overton was explicitly venturing into philosophical terrain. His obiter dicta are about as remote from well-founded opinion in the philosophy of science as Creationism is from respectable geology. It simply will not do for the defenders of science to invoke philosophy of science when it suits them (e.g., their much-loved principle of falsifiability comes directly from the philosopher Karl Popper) and to dismiss it as "arcane" and "remote" when it does not. However noble the motivation, bad philosophy makes for bad law.

The victory in the Arkansas case was hollow, for it was achieved only at the expense of perpetuating and canonizing a false stereotype of what science is and how it works. If it goes unchallenged by the scientific community, it will raise grave doubts about that community's intellectual integrity. No one familiar with the issues can really believe that anything important was settled through anachronistic efforts to revive a variety of discredited criteria for distinguishing between the scientific and the non-scientific. Fifty years ago, Clarence Darrow asked, à propos the Scopes trial, "Isn't it difficult to realize that a trial of this kind is possible in the twentieth century in the United States of America?" We can raise that question anew, with the added irony that, this time, the pro-science forces are defending a philosophy of science which is, in its way, every bit as outmoded as the "science" of the creationists.

Reference

1. "The Creationist Threat: Science Finally Awakens," VI *The Skeptical Inquirer* 3 (Spring 1982): 2–5.

Response to Laudan's Commentary: Pro Judice

Michael Ruse

As always, my friend Larry Laudan writes in an entertaining and provocative manner, but, in his complaint against Judge William Overton's ruling in *McLean v. Arkansas*, Laudan is hopelessly wide of the mark. Laudan's outrage centers on the criteria for the demarcation of science which Judge Overton adopted, and the judge's conclusion that, evaluated by these criteria, creation-science fails as science. I shall respond directly to this concern—after making three preliminary remarks.

First, although Judge Overton does not need defense from me or anyone else, as one who participated in the Arkansas trial, I must go on record as saying that I was enormously impressed by his handling of the case. His written judgment is a first-class piece of reasoning. With cause, many have criticized the State of Arkansas for passing the "Creation-Science Act," but we should not ignore that, to the state's credit, Judge Overton was born, raised, and educated in Arkansas.

Second, Judge Overton, like everyone else, was fully aware that proof that something is not science is not the same as proof that it is religion. The issue of what constitutes science arose because the creationists claim that their ideas qualify as genuine science rather than as fundamentalist religion. The attorneys developing the American Civil Liberties Union (ACLU) case believed it important to show that creation-science is not genuine science. Of course, this demonstration does raise the question of what creation-science really is. The plaintiffs claimed that creation-science always was (and still is) religion. The plaintiffs' lawyers went beyond the negative argument (against science) to make the positive case (for religion). They provided considerable evidence for the religious nature of creation-science, including such things as the creationists' explicit reliance on the Bible in their various writings. Such arguments seem about as strong as one could wish, and they were duly noted by Judge Overton and used in support of his ruling. It seems a little unfair, in the context, therefore, to accuse him of "specious" argumentation. He did not adopt the naïve dichotomy of "science or religion but nothing else."

Third, whatever the merits of the plaintiffs' case, the kinds of conclusions and strategies apparently favored by Laudan are simply not strong enough for legal purposes. His strategy would require arguing that creation-science is weak science and therefore ought not to be taught. He states:

The core issue is not whether Creationism satisfies some undemanding and highly controversial definitions of what is scientific; the real question is whether the existing evidence provides stronger arguments for evolutionary theory than for Creationism. Once that question is settled, we will know what belongs in the classroom and what does not.

Unfortunately, the U.S. Constitution does not bar the teaching of weak science. What it bars (through the Establishment Clause of the First Amendment) is the teaching of religion. The plaintiffs' tactic was to show that creation-science is less than weak or bad science. It is not science at all.

Turning now to the main issue, I see three questions that must be addressed. Using the five criteria listed by Judge Overton, can one distinguish science from non-science? Assuming a positive answer to the first question, does creation-science fail as genuine science when it is judged by these criteria? And, assuming a positive answer to the second, does the opinion in *McLean* make this case?

The first question has certainly tied philosophers of science in knots in recent years. Simple criteria that supposedly give a clear answer to every case—for example, Karl Popper's single stipulation of falsifiability[1]—will not do. Nevertheless, although there may be many grey areas, white does seem to be white and black does seem to be black. Less metaphorically, something like psychoanalytic theory may or may not be science, but there do appear to be clear-cut cases of real science and of real non-science. For instance, an explanation of the fact that my son has blue eyes, given that both parents have blue eyes, done in terms of dominant and recessive genes and with an appeal to Mendel's first law, is scientific. The Catholic doctrine of transubstantiation (i.e., that in the Mass the bread and wine turn into the body and blood of Christ) is not scientific.

Furthermore, the five cited criteria of demarcation do a good job of distinguishing the Mendelian example from the Catholic example. Law and explanation through law come into the first example. They do not enter the second. We can test the first example, rejecting it if necessary. In this sense, it is tentative, in that something empirical might change our minds. The case of transubstantiation is different. God may have His own laws, but neither scientist nor priest can tell us about those which turn bread and wine into flesh and blood. There is no explanation through law. No empirical evidence is pertinent to the miracle. Nor would the believer be swayed by any empirical facts. Microscopic examination of the Host is considered irrelevant. In this sense, the doctrine is certainly not tentative.

One pair of examples certainly do not make for a definitive case, but at least they do suggest that Judge Overton's criteria are not quite as irrelevant as Laudan's critique implies. What about the types of objections (to the criteria) that Laudan does or could make? As far as the use of law is concerned, he might complain that scientists themselves have certainly not always been that particular about reference to law. For instance, consider the following claim by Charles Lyell in his *Principles of Geology* (1830/3): "We are not, however, contending that a real departure from the antecedent course of physical events cannot be traced in the introduction of man."[2] All scholars agree that in this statement Lyell was going beyond law. The coming of man required special divine intervention. Yet, surely the *Principles* as a whole qualify as a contribution to science.

Two replies are open: either one agrees that the case of Lyell shows that science has sometimes mingled law with non-law; or one argues that Lyell (and others) mingled science and non-science (specifically, religion at this point). My inclination is to argue the latter. Insofar as Lyell acted as scientist, he appealed only to law. A century and a half ago, people were not as conscientious as today about separating science and religion. However, even if one argues the former alternative—that some science has allowed place for non-lawbound events—this hardly makes Laudan's case. Science, like most human cultural phenomena, has evolved. What was allowable in the early nineteenth century is not necessarily allowable in the late twentieth century. Specifically, science today does not break with law. And this is what counts for us. We want criteria of science for today, not for yesterday. (Before I am accused of making my case by fiat, let me challenge Laudan to find one point within the modern geological theory of plate tectonics where appeal is made to miracles, that is, to breaks with law. Of course, saying that science appeals to law is not asserting that we know all of the laws. But who said that we did? Not Judge Overton in his opinion.)

What about the criterion of tentativeness, which involves a willingness to test and reject if necessary? Laudan objects that real science is hardly all that tentative: "[H]istorical and sociological researches on science strongly suggest that the scientists of any epoch likewise regard some of their beliefs as so fundamental as not to be open to repudiation or negotiation."

It cannot be denied that scientists do sometimes—frequently—hang on to their views, even if not everything meshes precisely with the real world. Nevertheless, such tenacity can be exaggerated. Scientists, even Newtonians, have been known to change their minds. Although I would not want to say that the empirical evidence is all-decisive, it plays a major role in such mind changes. As an example, consider a major revolution of our time, namely, that which occurred in geology. When I was an undergraduate in 1960, students were taught that continents do not move. Ten years later, they were told that they do move. Where is the dogmatism here? Furthermore, it was the new empirical evidence—e.g., about the nature of the sea-bed—that persuaded geologists. In short, although science may not be as open-minded as Karl Popper thinks it is, it is not as close-minded as, say, Thomas Kuhn[3] thinks it is.

Let me move on to the second and third questions, the status of creation-science and Judge Overton's treatment of the problem. The slightest acquaintance with the creation-science literature and Creationism movement shows that creation-science fails abysmally as science. Again consider the passage, written by one of the leading creationists, Duane T. Gish, in *Evolution? The Fossils Say No!*:

Creation. By creation we mean the bringing into being by a supernatural Creator of the basic kinds of plants and animals by the process of sudden, or fiat, creation.

We do not know how the Creator created, what processes He used, *for He used processes which are not now operating anywhere in the natural universe.* This is why we refer to creation as Special Creation. We cannot discover by scientific investigations anything about the creative processes used by the Creator.[4]

The following similar passage was written by Henry M. Morris, who is considered to be the founder of the creation-science movement:

... it is ... quite impossible to determine anything about Creation through a study of present processes, because present processes are not created in character. If man wishes to know anything about Creation (the time of Creation, the duration of Creation, the order of Creation, the methods of Creation, or anything else) his sole source of true information is that of divine revelation. God was there when it happened. We were not there . . . therefore, we are completely limited to what God has seen fit to tell us, and this information is in His written Word. This is our textbook on the science of Creation![5]

By their own words, therefore, creation-scientists admit that they appeal to phenomena not covered or explicable by any laws that humans can grasp as laws. It is not simply that the pertinent laws are not yet known. Creative processes stand outside law as humans know it (or could know it) on earth—at least there is no way that scientists can know Mendel's laws through observation and experiment. Even if God did use His own laws, they are necessarily veiled from us forever in this life, because Genesis says nothing of them.

Furthermore, there is nothing tentative or empirically checkable about the central claims of creation-science. Creationists admit as much when they join the Creation Research Society (the leading organization of the movement). As a condition of membership applicants must sign a document specifying that they now believe and will continue to believe:

(1) The Bible is the written Word of God, and because we believe it to be inspired throughout, all of its assertions are historically and scientifically true in all of the original autographs. To the student of nature, this means that the account of origins in Genesis is a factual presentation of simple historical truths. (2) All basic types of living things, including man, were made by direct creative acts of God during Creation Week as described in Genesis. Whatever biological changes have occurred since Creation have accomplished only changes within the original created kinds. (3) The great Flood described in Genesis, commonly referred to as the Noachian Deluge, was an historical event, worldwide in its extent and effect. (4) Finally, we are an organization of Christian men of science, who accept Jesus Christ as our Lord and Savior. The account of the special creation of Adam and Eve as one man and one woman, and their subsequent fall into sin, is the basis for our belief in the necessity of a Savior for all mankind. Therefore, salvation can come only thru accepting Jesus Christ as our Savior.[6]

It is difficult to imagine evolutionists signing a comparable statement, that they will never deviate from the literal text of Charles Darwin's *On the Origin of Species*. The non-scientific nature of creation-science is evident for all to see, as is also its religious nature. Moreover, the quotes I have used above were all used by Judge Overton, in the *McLean* opinion, to make exactly the points I have just made. Creation-science is not genuine science, and Judge Overton showed this.

Finally, what about Laudan's claims that some parts of creation-science (e.g., claims about the Flood) are falsifiable and that other parts (e.g., about the originally created "kinds") are revisable? Such parts are not falsifiable or revisable in a way indicative of genuine science. Creation-science is not like physics, which exists as part of humanity's common cultural heritage and domain. It exists solely in the imaginations and writing of a relatively small group of people. Their publications (and stated intentions) show that, for example, there is no way they will relinquish belief in the Flood, whatever the evidence.[7] In this sense, their doctrines are truly unfalsifiable.

Furthermore, any revisions are not genuine revisions, but exploitations of the gross ambiguities in the creationists' own position. In the matter of origins, for example, some elasticity could be perceived in the creationist position, given the conflicting claims regarding the possibility of (degenerative) change within the originally created "kinds." Unfortunately, any open-mindedness soon proves illusory; for creationists have no real idea about what God is supposed to have created in the beginning, except that man was a separate species. They rely solely on the Book of Genesis:

And God said, Let the waters bring forth abundantly the moving creature that hath life, and the fowl that may fly above the earth in the open firmament of heaven.

And God created great whales, and every living creature that moveth, which the waters brought forth abundantly, after their kind, and every winged fowl after his kind: and God saw that it was good.

And God blessed them, saying Be fruitful, and multiply, and fill the waters in the seas, and let fowl multiply in the earth.

And the evening and the morning were the fifth day.

And God said, Let the earth bring forth the living creature after his kind, cattle, and creeping thing, and beast of the earth after his kind: and it was so.

And God made the beast of the earth after his kind, and cattle after their kind, and everything that creepeth upon the earth after his kind: and God saw that it was good.[8]

But the *definition* of "kind," what it really is, leaves creationists as mystified as it does evolutionists. For example, creationist Duane Gish makes this statement on the subject:

[W]e have defined a basic kind as including all of those variants which have been derived from a single stock . . . We cannot always be sure, however, what constitutes a separate kind. The division into kinds is easier the more the divergence observed. It is obvious, for example, that among invertebrates the protozoa, sponges, jellyfish, worms, snails, trilobites, lobsters, and bees are all different kinds. Among the vertebrates, the fishes, amphibians, reptiles, birds, and mammals are obviously different basic kinds.

Among the reptiles, the turtles, crocodiles, dinosaurs, pterosaurs (flying reptiles), and ichthyosaurs (aquatic reptiles) would be placed in different kinds. Each one of these major groups of reptiles could be further subdivided into the basic kinds within each.

Within the mammalian class, duck-billed platypus, bats, hedgehogs, rats, rabbits, dogs, cats, lemurs, monkeys, apes, and men are easily assignable to different basic kinds. Among the apes, the gibbons, orangutans, chimpanzees, and gorillas would each be included in a different basic kind.[9]

Apparently, a "kind" can be anything from humans (one species) to trilobites (literally thousands of species). The term is flabby to the point of inconsistency. Because humans are mammals, if one claims (as creationists do) that evolution can occur within but not across kinds, then humans could have evolved from common mammalian stock—but because humans themselves are kinds such evolution is impossible.

In brief, there is no true resemblance between the creationists' treatment of their concept of "kind" and the openness expected of scientists. Nothing can be said in favor of creation-science or its inventors. Overton's judgment emerges unscathed by Laudan's complaints.

References

1. Karl Popper, *The Logic of Scientific Discovery* (London: Hutchinson, 1959).

2. Charles Lyell, *Principles of Geology*, Volume I (London: John Murray, 1830), p. 162.

3. Thomas Kuhn, *The Structure of Scientific Revolutions* (Chicago: University of Chicago Press, 1962).

4. Duane Gish, *Evolution? The Fossils Say No!*, 3rd edition (San Diego: Creation-Life Publishers, 1979), p. 40 (his italics).

5. Henry M. Morris, *Studies in the Bible and Science* (Philadelphia: Presbyterian and Reformed Publishing Company, 1966), p. 114.

6. Application form for the Creation Research Society, reprinted in *McLean v. Arkansas*, footnote 7.

7. See, for instance, Henry M. Morris, *Scientific Creationism* (San Diego: Creation-Life Publishers, 1974); and my own detailed discussion in Michael Ruse, *Darwinism Defended: A Guide to the Evolution Controversies* (Reading, MA: Addison-Wesley, 1982).

8. Genesis, Book I, Verses 20–25.

9. Gish, *op. cit.,* pp. 34–35.

Creationism and Education in the Physical Sciences*   Stephen G. Brush

What would happen to science education if "creation-science" laws such as those passed in Arkansas and Louisiana were put into effect? In this essay I explore, from the perspective of a historian of science—whose formal education in biology ended with a high-school course more than thirty years ago—the possible consequences of the success of creationists' claims, and discuss how some of their claims are related to the history and philosophy of science.

Definition of Creationism—A Young Earth

Arkansas Act 590 defined "creation-science" as including in particular "a relatively recent inception of the earth and living kinds." The Louisiana Law (Act 685) does not specify what is meant by "creation-science"; presumably this would be done by the seven creation-scientists to be appointed by the state governor. In practice, however, when a state or local school board has mandated that Creationism (or creation-science) be taught, the materials produced by the Institute for Creation Research and similar groups have been used. So, for the discussion in this essay, I will rely primarily on Arkansas Act 590 and on Henry Morris' book *Scientific Creationism*, the reference most frequently cited by creationists.[1] Morris and his colleagues claim that the universe was created as recently as 6,000 to 10,000 years ago. Although it is possible to reject evolution without accepting the "young-earth" version of Cre-

*This essay is based on an address delivered to the 57th Annual Meeting of the West Virginia Academy of Science, to be published in somewhat different form in the *Proceedings* of the Academy.

ationism, that is the version most likely to be taught in public schools if the "equal time" or "balanced treatment" demand is enforced. What, then, might be the consequences of requiring science teachers to present arguments for a young earth?

Consequences in the Science Classroom

Earth Sciences
Although most of the public controversy has centered on the teaching of Darwinian evolution, strict interpretation of the Arkansas law (or of similar legislation) would have a greater impact on earth-science and astronomy courses than on biology courses because "evolution-science" is defined to include generally-accepted theories in geology and astronomy [Act 590, Section 4(a), items 1, 5, and 6]. In a biology text, the discussion of evolution can be restricted to a single chapter (even though this may render the rest of the book uninteresting), but the same cannot be done with earth science and astronomy texts. All of historical geology, plate tectonics, and theories of the early development of the earth depend on a time scale established by radioactive dating that is rejected by creationists. According to Rick Duschl of the University of Maryland, an educator familiar with earth-science curricula, more than 60 percent of the material in a typical textbook would be defined as "evolution-science" and thus required to be balanced by creationist material.[2] This means that, if the model creationist legislation (or its principles) were adopted, at least 30 percent of the present content of such a course would have to be eliminated to make room for material that most scientists consider to be of no scientific value.

Astronomy
Creationists also call into question the very existence of stars that are millions of light years away. If the universe was created no more than 10,000 years ago, how can we see stars that are millions of light years away from us, they ask. Wouldn't they have had to exist millions of years ago to send out the light which reaches us now? For creationists such as Henry Morris and Duane Gish, this paradox is easily resolved: they believe the stars were created for the purpose of being "signs" to man. God could easily have created their light in space en route to earth in such a way that the stars appear to be very old:

Whatever is truly created—that is, called instantly into existence out of nothing—must certainly look as though it had been there prior to its creation. Thus it has an appearance of age.[3]

The creationist explanation of distant stars, in fact, is a modern version of Philip Henry Gosse's "omphalos" (navel) theory; that is, Adam and Eve were created with navels to make it appear that they were born in the usual way.[4] But such an approach creates an immediate problem for the science teacher required to give equal time to Creationism. How can the omphalos explanation be presented without mentioning the Creator who "planted" the abundant evidence of the antiquity of the earth and the rest of the universe?

Creationists reject the big-bang cosmology, but their criticism completely ignores the difficulties that have bothered some recent investigators. Instead, they attack the most spectacular success of the big-bang theory—the prediction of the black-body radiation that permeates space—by arguing that the radiation produced by the primeval fireball would have escaped by crossing the boundary of the expanding universe. Of course anyone with a rudimentary knowledge of Einstein's theory of relativity (on which the big-bang theory is based) realizes that there is no such boundary.[5] Nevertheless, teaching creationist astronomy will require a new version of relativity theory, something that creationists are currently developing.[6]

Physics

By claiming that the second law of thermodynamics forbids the development of complex forms of life by natural processes, creationists have attempted to use physics against evolution. In truth, they not only ignore the generally adopted version of the second law, but are also quite willing, when it suits their purposes, to discard other well-established physical theories and propose new ones in their place.

According to Morris, the second law states that "every system left to its own devices tends to move from order to disorder."[7] This oversimplified version, as Joel Cracraft points out elsewhere in this volume, ignores important qualifications. The system must be closed to the flow of matter and energy; otherwise the law in its original form does not apply at all. For a system in thermodynamic equilibrium, the most stable state is *not* the one with the highest disorder or entropy, except at very high temperatures; it is the one with the lowest *free energy F* $= E - TS$ (where E = energy, T = temperature, S = entropy). As one lowers the temperature, the state with the lowest energy (usually an ordered crystalline arrangement of the molecules) will be favored. Thus, the formation of snowflakes from water vapor, a familiar natural process which involves a transition from disorder to order, is perfectly consistent with the second law. In general, interatomic forces can result in the spontaneous formation of complicated molecules (including organic molecules) under appropriate conditions.

The most flagrant contradiction of Creationism and physics arises from the attempt to evade the results of radiometric dating. Since I have discussed this problem in considerable detail elsewhere, I will mention only a few highlights.[8] First, creationists claim that radioactive decay constants might have varied enough in the past to produce an error of a factor of a million in the estimated age of the earth. (They say the earth, as well as the rest of the universe, is only a few *thousand* years old, while scientists say it is a few *billion*.) Yet their alleged evidence for such variation collapses under close examination. Second, to explain how such a variation in decay rate could have occurred, creationists rely on a theory by H. C. Dudley which postulates that radioactive decay is not a random process but depends on the coupling of the nucleus to a surrounding ethereal medium. As Dudley himself admits, this theory requires us not only to reject Einstein's theory of relativity (in order to revive a nineteenth-century ether) but also to postulate "hidden variables" at a subquantum level, contrary to the principles of quantum mechanics.

Impact on Higher Education and Curriculum Development

To save their theory and shore up their objections to evolution, creationists reject or distort the basic principles of astronomy, relativity, thermodynamics, nuclear physics, quantum mechanics, and geophysics. And it is reasonable to predict that if creationists' biology is adopted, then they will demand equal time in physics classes, astronomy classes, and so forth. If any state passed a law requiring the teaching of creation-science in public schools, then the universities and other teacher-training institutions in that state would also be expected to offer courses in "creationist astronomy" and "creationist physics" as well as "creationist geology" and "creationist biology."

In my opinion the real thrust of the creationist movement is not so much to put Creationism into the schools as to take evolution out. Unless they control the public schools completely or can daily monitor classrooms, creationists would not benefit by having their doctrines taught by skeptical teachers to skeptical students. I suspect that most students, for example, would realize that omphalos explanations of phenomena such as stars appearing to be millions of light years away are simply inappropriate for science. If such arguments were presented, students would quickly lose all respect for Creationism.

A more likely result of the present controversy is that many schools will teach neither creation-science nor evolution—a reaction that could push science education back to its status following the Scopes trial, when evolution was almost completely eliminated from U.S. high-school biology courses (until restored in the 1960s). Biology was taught as a collection of facts and descriptions to be memorized with no theory to provide a unifying explanatory framework. I was a victim of this prohibition—my high school biology course was so boring that I never wanted to take any more courses in the subject. If textbook publishers and school boards decide to play it safe by playing down evolution,[9] there is a real danger that biology courses could be altered even without legislation.

Consequences for the Philosophy of Science

Act 590 includes the phrase "evolution cannot be experimentally observed, fully verified, or logically falsified" [Section 7(c)]. This statement reflects two standard creationist arguments, both fallacious but for rather different reasons. The first is that one can distinguish between "microevolution," which can be experimentally observed (e.g., industrial melanism, development of strains of organisms resistant to antibiotics, and so forth), and "macroevolution," which takes place over a long period of time and therefore cannot be directly observed. The creationists then incorporate microevolution into their own model [Act 590, Section 4(a)(3)] although they define "evolution-science" as including only macroevolution, not microevolution [Section 4(b)(3)]. Of course, scientists need not accept this peculiar definition of evolution. If creationists admit that evolution works to a small degree, the burden of proof is on them to show that it suddenly stops working as some "fixed limit."

The second argument is that the theory of evolution cannot be logically falsified because it merely interprets events that have happened in the distant past rather than attempting to predict what will happen in the future. This objection goes back to the doctrine made popular by Karl Popper, that a theory is not scientific unless it makes predictions that can be tested and possibly found incorrect. Popper wanted to establish a criterion of demarcation between science and pseudo-science. Theories in a pseudo-science (he mentioned Marxism and psychoanalysis as specific examples) are so flexible that they can explain anything, hence they can never be tested. Popper himself once argued that Darwinian evolutionary theory is not a falsifiable hypothesis but a "metaphysical research programme."[10] Later he recanted this view and decided that Darwin's theory is indeed a legitimate scientific hypothesis, but creationists ignore his change of mind.[11]

Popper's original criterion defined "prediction" so narrowly that not only evolutionary biology but geology and astronomy would be excluded from the sciences. Any attempt to deal with phenomena that take place over a large domain of space or time, and that therefore cannot be brought into the laboratory for controlled tests, would have to be called pseudo-science since it could not generate predictions of specific future events. Because we do generally consider geology and astronomy to be sciences, we must either broaden the concept of "prediction" or use another criterion for "scientific."

Testability of Theories of Origin

Several good treatments of the scientific status of evolution are available. Two of the most recent, by Michael Ruse and by Philip Kitcher, analyze creationists' objections to evolution and show that they are groundless.[12] Leaving aside the testability of biological theories, however, what about the creationists' more general claim that any theory of *origins* is in principle untestable and therefore not scientific? This claim provides the rationale for their argument that because neither Creationism nor evolution (in the broad sense defined in Act 590) can be proved or disproved, both should be presented. Students should be allowed to choose.

Although the claim sounds plausible at first, it fails because two well-known theories of origins have, in fact, been tested: one was dramatically confirmed, the other refuted (or at least abandoned by its major advocate). Creationists reject both theories without giving serious consideration to the evidence for or against them.

The first example is the big-bang theory of the origin of the universe, as developed in the 1940s by George Gamow and his students. They predicted that the electromagnetic radiation produced in the original explosion (now called the "primeval fireball") would have cooled off over a period of several billion years and should now be present as background radiation at a temperature of a few degrees above absolute zero. A. A. Penzias and R. W. Wilson confirmed this prediction in 1965, and it is generally regarded as a crucial piece of evidence in favor of the big-bang cosmology. As noted above, creationists can deal with this success only be rejecting (or misconstruing) Einstein's model of a curved space-time that is finite but has no boundaries.

The second example is the theory of the origin of the moon advocated by Harold Urey and others in the 1960s. Urey argued that the moon was formed elsewhere in the solar system and captured by the earth, and that it has preserved on its surface a record of events that occurred billions of years ago. This, in his view, was the major justification for an extensive program of manned lunar exploration—that we could learn something about early solar-system history. But the Apollo program produced evidence that contradicted the detailed predictions of Urey's theory, and he eventually abandoned it in favor of the idea that the moon was spun off from the earth in a fission-like process.[13] Creationists also reject the capture theory, but seem unaware of the valid scientific reasons for doing so.[14]

Many scientists regard Creationism as unscientific because it is untestable—a view reflected in Judge Overton's decision in *McLean v. Arkansas*. At the same time, they argue that the major tenets of Creationism have been refuted. Is this inconsistent? Not if one distinguishes between Creationism as a set of propositions and Creationism as an activity of particular people. Creationism *was* a science in the eighteenth century, but it was decisively refuted in the nineteenth century, and I would argue that its twentieth-century advocates behave unscientifically when they ignore the refuting evidence. In other words, Creationism is testable, has been tested, and has been refuted. It can only be advocated now as a pseudo-science.

Nevertheless, creationists have one other opportunity to show that their doctrine is a testable scientific hypothesis. Frank Tipler of Tulane University has worked out a consistent cosmological theory based on the hypothesis that the universe began in 4004 B.C. with the explosion of numerous black holes.[15] Tipler's theory is also a variant of the omphalos theory, but one which has testable consequences. In particular, it says that if we look about 6,000 years ago into space we should see some evidence of black-hole explosions that created the universe. About 10^6 black-hole explosions per second must be occurring at 6,000 light years (the number of explosions in the direction of the galactic center being about twice the number in the opposite direction). Furthermore, the theory asserts that no black-hole explosion can occur closer to the earth than 6,000 light years.

Because Tipler makes fairly definite numerical predictions, we could test his theory by simply looking for the predicted explosions—if theorists knew exactly what a black-hole explosion looked like. Unfortunately, calculating what the explosion would look like involves understanding elementary-particle physics at energies unreachable by present accelerators. Future research in elementary particle physics may enable us to perform a definite test, but we only guess now that we will need observations with a very sensitive gamma-ray telescope, probably placed in a space shuttle. Testing the theory could thus be very expensive, a price that believers in "creation-science" should be willing to pay to show that their theory is a testable scientific hypothesis.*

Consequences for Historians of Science

I am sometimes asked why I bother with the creation-evolution controversy; some scientists consider it a waste of time to argue with people who will never change their opinions in the face of contrary evidence. One answer might be that my own field, the history of science, is frequently distorted to provide arguments for Creationism. Embarrassed by the fact that no reputable modern scientists support their theories, for example, creationists claim that great scientists of the past were creationists. Morris recently published a list of scientists who had founded various disciplines and made great discoveries, and were allegedly creationists. The British physicist William Thomson, Lord Kelvin, appears on this list four times.[16] But Lord Kelvin actually repudiated Creationism; he was, in modern terms, a "theistic evolutionist."[17]

A more substantial reason for my interest in Creationism is that it represents a reincarnation of a pre-modern world view. Any historian would welcome the opportunity to travel by time machine to the trial of Galileo—not so much to check on controversial details of the proceedings as to observe the behavior of men who, in the face of all contrary evidence, stick to what they consider a literal reading of the Bible. Nowadays, few individuals will seriously maintain that the earth is the center of the universe, yet Creationism is part of the same outlook. This modern controversy allows historians to view a "rerun" of the

*Some aspects of the theory should be appealing to the Moral Majority: it relies on the principle of "cosmic censorship" that prevents anyone from observing the "naked singularities" resulting from black-hole explosions.—SB

1860 debate between Bishop Wilberforce and T. H. Huxley, and to observe creationist speakers and attempt to analyze their effectiveness.

It is a challenging problem for historians to explain why Creationism has been revived at this particular time and has gained substantial public support. (For example, a recent Gallup poll found that 44 percent of all Americans polled believed that "God created man pretty much in his present form at one time during the last 10,000 years," 38 percent believed that "man has developed over millions of years from less advanced forms of life, but God guided this process including man's creation," and 9 percent that it took millions of years but that "God had no part in this process.")[18] My view is that the movement is cor-related with a complex of ideas and values which might be identified with a generalized cultural romanticism—a conservative, holistic view-point that is opposed to individualism and scientific rationality, seems always to be present, but is enunciated only every fifty or sixty years.[19] The creationists consider not only evolution but also "secular humanism to be their enemy." The following is a typical definition by Kelly Segraves, Director of the Creation Science Research Center:

Humanism is a far-reaching social program that aims for the estab-lishment throughout the world of democracy (lowest common denom-inator mob rule), peace and a high standard of living. This developing program of the humanists is, they say, ever open to experimental testing, newly discovered facts and rigorous reasoning.

All of this is anathema to the creationist whose philosophy is "dia-metrically opposed, mutually exclusive, with respect to life and the value of life."[20]

It is not surprising to me that the 1970s revival of political con-servatism is associated with religious fundamentalism and Creationism. Much of the progress in civil rights, environmental protection, and social welfare made before 1970 has been shown to be reversible when the political climate changes. Is science exempt from similar reaction?

Fortunately, history is not completely cyclic. Society will not revert to the eighteenth century (or even to 1925) in its thinking about the human condition. And I am convinced that Creationism will again be defeated (although not destroyed) as sterile dogmatism proves to be no match for open-minded scientific inquiry.

Notes and References

1. Henry Morris, ed. *Scientific Creationism* (San Diego: Creation-Life Publishers, 1974).

2. Duschl's estimate is based on *Investigating the Earth* by Matthews *et al.,* third edition (Boston: Houghton Mifflin, 1978), and the teacher's manual that outlines a teaching schedule for this text, an outgrowth of the original Earth Science Curriculum Project sponsored by the American Geological Institute.

3. Henry Morris, *The Remarkable Birth of Planet Earth* (Minneapolis: Dimension Books, 1972), p. 62. The statement by Duane Gish was made in "Creation vs. Evolution—Battle in the Classroom," a television film on the creation-evolution controversy, produced by KPBS-TV (San Diego) and broadcast on WETA (Washington, D.C.) on 14 July 1982. (Transcript available for $3.00. Write: "Creation," KPBS-TV, San Diego, CA 92182.)

4. See the discussion by F. C. Haber, *The Age of the World, Moses to Darwin* (Baltimore: Johns Hopkins Press, 1959), pp. 246–250; and by Martin Gardner, *Fads and Fallacies in the Name of Science,* second edition (New York: Dover, 1957), pp. 124–127.

5. Russell Akridge, Thomas Barnes, and Harold S. Slusher, "A Recent Creation Explanation of the 3°K Background Black Body Radiation," 18 *Creation Research Society Quarterly* (1981): 159–162.

6. See, for example, T. G. Barnes and R. J. Upham, Jr., "Another Theory of Gravitation: An Alternative to Einstein's General Theory of Relativity," 12 *Creation Research Society Quarterly* (1976): 194–197.

7. Morris (1974), *op. cit.*, p. 25.

8. Stephen G. Brush, "Finding the Age of the Earth: By Physics or By Faith?" 30 *Journal of Geological Education* 1 (January 1982): 34–58. The arguments given in this paper are sufficient to refute the latest creationist attack on radiometric dating as described in Theodore W. Rybka, "Consequences of Time Dependent Nuclear Decay Indices on Half Lives," *Impact* 106 (San Diego: Institute for Creation Research, April 1982). To supplement my discussion of T. G. Barnes' estimate of the age of the earth from the decay of its magnetic field, see G. Brent Dalrymple, "Can the Earth Be Dated from Decay of the Magnetic Field?" (Menlo Park: U.S. Geological Survey, preprint 10 August 1982), to be published in *Journal of Geological Education.* Also see Brent Dalrymple, "How Old Is the Earth? A Reply to 'Scientific Creationism'," presented at an American Association for the Advancement of Science symposium, Santa Barbara (June 1982), to be published in the proceedings of that symposium.

9. Gerald Skoog, "Topic of Evolution in Secondary School Biology Textbooks: 1900–1977," 63 *Science Education* (1979): 621–640. Stephen Jay Gould writes about the textbooks used in the 1950s: "Thus were millions of children deprived of their chance to study one of the most exciting and influential ideas in science, the central theme of all biology. A few hundred, myself included, possessed the internal motivation to transcend this mockery of education. . . ." 91 *Natural History* 3 (March 1982): 4–10.

10. Karl Popper, *Unended Quest* (La Salle: Open Court, 1976), pp. 167–180.

11. Karl Popper, "Natural Selection and the Emergence of Mind," 32 *Dialectica* (1978): 339–355; "Evolution," 87 *New Scientist* (1980): 611. Robert E. Kofahl and H. Zeisel, "Popper on Darwinism," 212 *Science* (1981): 873. William J. Board, "Creationists Limit Scope of Evolution Case," 211 *Science* (20 March 1981): 1331–1332.

12. Michael Ruse, *Darwinism Defended: A Guide to the Evolution Controversies* (Reading, MA: Addison-Wesley, 1982); Philip Kitcher, *Abusing Science* (Cambridge, MA: MIT Press, 1982).

13. Stephen G. Brush, "Nickel for Your Thoughts: Urey and the Origin of the Moon," 217 *Science* (3 September 1982): 891–898.

14. John C. Whitcomb and Donald B. De Young, *The Moon, Its Creation, Form, and Significance* (Winona Lake, IN: BMH Books, 1978).

15. Frank J. Tipler, "Did the Universe Begin in 4004 B.C.?" (New Orleans: Departments of Mathematics and Physics, Tulane University, 1982), preprint.

16. Henry Morris, "Bible-Believing Scientists of the Past," *Impact* 103 (San Diego: Institute for Creation Research, January 1982); see also his *Men of Science, Men of God* (San Diego: Creation-Life Publishers, 1982), pp. 85–87.

17. Address to the British Association for the Advancement of Science, 1871; reprinted in Kelvin's *Popular Lectures and Addresses*, and extracted in *Victorian Science*, edited by George Basalla, William Coleman, and Robert H. Kargon (Garden City, NY: Anchor Books, 1970), pp. 101–128. Also see Stephen G. Brush, "Kelvin Was Not a Creationist," 8 *Creation/Evolution* (1982): 11–14.

18. Gallup Poll, 23–26 July 1982 (Princeton: Gallup Organization, 1982); George Gallup, "Public Evenly Divided Between Evolutionists, Creationists," Gallup Poll press release, 29 August 1982.

19. Stephen G. Brush, *The Temperature of History: Phases of Science and Culture in the Nineteenth Century* (New York: Burt Franklin, 1978); also, "The Chimerical Cat: Philosophy of Quantum Mechanics in Historical Perspective," 10 *Social Studies of Science* (1980): 393–447. In case anyone still doubts the antiscientific character of the current conservative movement (apart from its support for Creationism), I would point out that the Conservative Caucus and the National Conservative Political Action Committee not only want to cut back sharply on social programs but also propose to cut the budget of the National Institutes of Health by 50 percent, and eliminate the National Science Foundation entirely! See Paul Taylor, "Conservatives Offer Their Own Budget," *Washington Post* (17 April 1982): A8.

20. "A Revolution Against Humanism," *Creation-Science Report* (January 1980): 2.

Some Comments on Biology, Fundamentalism, and Education

Ann Rebecca Bleefeld

Socrates . . . Tell me: is it not true that learning about something means becoming wiser in that matter?

Theaetetus Of course.

Socrates And what makes people wise is wisdom, I suppose.

Theaetetus Yes.

Socrates And is that in any way different from knowledge?

Theaetetus Is what different?

Socrates Wisdom. Are not people wise in the things in which they have knowledge?

Theaetetus Certainly.

Socrates Then knowledge and wisdom are the same thing?

Theaetetus Yes.[1]

Sometimes when I think about students and teachers, the dialogue between Socrates and Theaetetus comes to mind as an ancient epitome of letting go of long-held illusions. As students, we try to grasp ideas and learn facts, hoping that in some way they will add something to our lives and add meaning to the world, render it concrete, defineable, malleable. It was difficult for Theaetetus to learn that this world is not so easily defined or within his control.

When the "Creationism" legislation and litigation began to be reported in the news, many of my colleagues and friends in New York thought it silly that religious ideas about the origin of life on earth could still be debated seriously anywhere in the United States. Others were annoyed that proponents of Creationism could presume to invade precious classroom time with "irrelevant" and "untruthful" notions about topics in biology. In general, however, most were complacent because they believed that elements of Fundamentalism and religious movements advocating "balanced treatment" in science education were, at least in New York, not part of the social, economic, political, and intellectual milieu within which educational decisions were made. When attempts were made in California and Kentucky to enforce teaching of Biblical interpretations of biological history in public schools, the defeat of such proposals lent credence to the skeptics' belief that such issues were not significant. I believe it has been shown, however, that events that affect the education process anywhere will sooner or later be felt by all educators and students.

A short time after the California and Kentucky issues had been fought, I was a graduate student and instructor in the Department of Biology at Queens College of the City University of New York. It was my job to lecture about and reiterate topics covered in regular lecture sessions, as well as to clarify problems concerning course material. During much of the semester, I described zoological life on earth, exploring scientific ideas, both historical and current, about relationships among animals and the reasons for their wide variety of forms. Several of my students openly challenged my approach, the scientific explanations I discussed, and my ability to reason in light of what they interpreted as my personal beliefs in the face of the Biblical account of creation. I suggested that, for those who wished to discuss this subject further, an informal session could be held outside of class. Later I questioned that decision and wondered if the whole class might not have benefited from an interchange of ideas on the subject. The sincere challenge made by those students called into question not only my reasons for having certain opinions, but also my role in the classroom, and my professional obligations to the institution at which I was both student and teacher. Naturally, I felt that I was obligated to adhere to course material that students would be required to learn. And yet I did not feel that Creationism was something to be ignored. The curriculum was set by the department and I needed more time to cover issues that I thought were justifiably raised by my students.

One of these students pursued the issue in conversation with me later, expressing feelings of resentment because I "neglected" Biblical explanations of biological phenomena. He seemed unable to accept the fact that I *chose* "evolutionism" rather than Fundamentalism as a means of reconstructing historical biological events. I asked him how *he* came to know things about the world, and he told me that the Bible was *the* sole authority on all things. I asked him if he ever saw discrepancies between his experience and Biblical pronouncements. "Of course," he answered, elaborating by saying that human beings are not able to understand everything that happens and therefore "we consult the Bible for explanations." "And when no explanation is given?" I asked. He answered, "We accept on faith, that God has reasons for things." "To which we are not privy," I said. "Yes," he agreed.

Sharing my recognition that life is often confusing did not seem to impress upon him the similarity rather than the disparity between us. He insisted that I was "looking in the wrong place for answers." To my student, the authority of the Bible and its accounts of biological phenomena had to be included in any attempt to discover the course of history of life on earth. For him, Biblical interpretations were not extraneous. I believe that whether my hypotheses stand or fall is affected by my earlier judgment of whether certain facts or other hypotheses are extraneous. But it was not clear to *me* why my belief that extraneous hypotheses (i.e., those not derivable from a study) should be omitted from scientific method was a choice unacceptable to him, even as a matter arising out of the judgment of another human being. During the rest of the semester, the student continued to grapple with a presentation of course material that was antithetical to his own convictions.

For me, there is something unsettling about a student, sitting day after day in a science classroom, who feels uncomfortable, unhappy, and even neglected by a standard presentation of course material. Usually accompanying this discomfort is the student's refusal to admit that his or her motivations in having given over his individual incentive to an authoritative dogma could be called into question. Biblical worth was never at issue in my confrontation with students in that class; rather, I was struggling with the ways in which students and teachers can make choices. Most distressing to me was the failure to find common ground on which to hold discussion and my inability to reassure those students that self-inspection need not further self-doubt or lead to recrimination.

Socrates wrestled with the doctrine that one can maintain a world view and not desire to scrutinize it, even if one's own position may be strengthened as a result. For Socrates, this self-examination was imperative, even if one's view is adopted from the teaching of someone else—and especially in that case. Dialogue that traces and maps the course of thought is no less essential for students and teachers today. Regardless of what happens in the courts, if educators refuse to reveal the assumptions, beliefs, and proofs that form the basis of their work, they will discourage open debate and youthful inquisitiveness. A door leading to further discovery will be closed.

As science teachers, we stand before our students and talk, describe, and explain, often without knowing why this is such a good thing to do. We hold scientific beliefs about the world without always analyzing why or noticing that our students may see things just a bit differently. Like Theaetetus, we are ever students in need of revelation.

Reference

1. Francis M. Cornford, trans., *Plato's Theory of Knowledge: The Theaetetus and the Sophist of Plato* (New York: Bobbs-Merrill, 1957), p. 20.

Creationism in the News: Marcel C. La Follette
Mass Media Coverage of
the Arkansas Trial

Because the controversy over Creationism in the public schools involves legal, religious, and scientific issues in equal measure, a review of the news coverage of an event such as the Arkansas "creation-science" trial points to some essential questions about the public communication of science. When science is intertwined with a social or political issue — and therefore not labeled as "science news" — how are the technical aspects covered and what can that coverage tell about how the mass media assess science? How extensive was news coverage of the *McLean* case — and why? Was it covered by science reporters, or primarily by political or religion writers? What emphases or biases are obvious, especially in the treatment of witness testimony?

This chapter reviews some of the news coverage of the Arkansas trial, and related creationist activities, in national newspapers and magazines, in the local Arkansas papers, and in the scientific press, from the time Act 590 was passed (March 1981) until after Judge William R. Overton's decision was handed down (January 1982). It also takes a quick look at television coverage, comparing how two public television documentaries treated the subject of creation-science. Although I do offer a few explanations and speculations in the last section, the chapter is intended primarily to present information that may be of use in future research on Creationism and to point out aspects of the news coverage which appear to call into question some traditional assumptions about the journalistic presentation of science.

A Re-Creation of Scopes?

The word "embarrassment" appeared repeatedly in the Arkansas newspapers throughout the trial, in news stories as well as in letters to the editor and in editorial comment. Some of this feeling centered on the circumstances of the bill's passage, although the consensus was that there the damage had already been done. One paper explained, "it's a little late to save face for Arkansas."[1] But there was also fear that the trial itself might become "Scopes II."

Most of the concern was fed by the "ballyhoo" reputation surrounding the 1925 trial of John T. Scopes, but these concerns were misplaced, for the Scopes trial actually had cultivated more journalistic excitement than civil celebration or riot. As Michael Williams wrote at the time in *Commonweal*:

The press has made this story. Its spotlight has been turned upon Dayton [Tennessee] as if by a common agreement among all editors everywhere that it was naturally the thing to do . . . no crowd of visitors has appeared. . . . The expected onrush of tourists and interested on-lookers has not materialized.[2]

In addition to several hundred reporters from newspapers such as the *New York Times, Baltimore Sun, Chicago Tribune*, the London *Times*, and *Philadelphia Inquirer*, many well-known writers of the day were present at Dayton, for example, H. L. Mencken, Joseph Wood Krutch, Nunnally Johnson of the Brooklyn *Eagle*, and Adolph Shelby Ochs of the *Chattanooga Times*. Scopes recalls in his autobiography that, by the time the trial opened, there were twenty-two Western Union operators "in a room off a grocery store," sending copy around the world.[3] Chicago radio station WGN spent $1000 per day in wire charges, and broadcast the final portions of the trial.[4] Almost a thousand people crammed into the hot courtroom; newsmen occupied a large portion of the seats, and were perched in windows or in any place they could find.[5] Michael Williams observed,

Fantastic as are many of the aspects of this amazing drama, ludicrous even as are some of the characters playing their parts amid movie cameras, buzzing aeroplanes, radio installations, clicking telegraph keys, chattering typewriters, and all the apparatus of up-to-date last-minute publicity, some quality deeper and graver and more disturbing than all the issues of ordinary life pervades everything said or done.[6]

In contrast to that bustle and drama, the 1981 Arkansas creation-science trial was calm and businesslike, something that won praise for Judge Overton from both press and counsel.[7]

The demeanor of the Arkansas trial reflects changes both in the attitudes of the judiciary toward the press and in styles of reporting since the time of Scopes. As is the rule in his Federal district, Judge

Overton did not allow cameras or recording devices in the courtroom; only sketch artists.[8] Seventy-one seats were set aside for the press inside the rail (where the jury normally sits[9]). No broadcast announcers softly descanted a play-by-play description of the testimony, and only two or three groups of demonstrators appeared, all quite sedate.

Even though the lack of showmanship and celebrities may have disappointed those who anticipated a "circus," the case did attract substantial national and international media attention. Seventy-five publications or news organizations registered more than one hundred journalists or production-crew members with Charles Gray, the U.S. Marshall in charge of press relations.[10] The local representatives included not only major state newspapers and broadcast stations but also reporters from the *Pea Ridge* (Ark.) *County Times*, the Little Rock Central High School *Tiger*, and the *Southern Mediator*, a black-oriented newspaper. In addition to U.S. wire services and broadcasting networks, crews were sent by Canadian and British broadcasting organizations. The popular science magazines (e.g., *Science 81, Discovery, Science News*), science news press (e.g., *Chemical & Engineering News, Science*), and religious press (e.g., *Christianity Today*) were represented in Little Rock, as were most of the same newspapers—the *New York Times, Baltimore Sun*, and *Philadelphia Inquirer*, for instance—that had sent reporters to cover the Scopes trial. After it was all over, U.S. Marshall Gray remarked: "we had no problems . . . everything went very smooth, and the press gave us no problems whatsoever."[11]

National Newspapers and Magazines[12]

Until Spring 1981, most U.S. newspapers had given only sporadic attention to Creationism or creation-science. Astute readers could have picked up hints of a growing confrontation—for example, the *New York Times* ran an article, "Foes of Evolution Theory Ask Equal School Time" (7 April 1980), that reported on a "sophisticated nationwide campaign" to persuade schools to revise their biology curricula—but most news about creationists consisted of short accounts of legal skirmishes in school boards or state legislatures.

In early March 1981, some U.S. newspapers had been covering a legal challenge, by creationists, to California textbook guidelines. Therefore, on March 19, when Act 590 was passed in Arkansas, it fitted in with a continuing *national* story; several papers—the *Los Angeles Times*, for example—ran it on the front page. From that time forward, news coverage of Creationism in general increased, as the level of legal activity fueled media interest.

Judged by traditional criteria for the importance of a news story—location in governmental hierarchies (Federal court), impact on the national interest, impact on large numbers of people, and significance for the past and future—the Arkansas case ranks high. By June 1981, the *New York Times* had surpassed the total number of stories it had run on the subject in the previous year. Many papers ran analyses of the renewed activities of the creationists—for example, front page articles appeared in the *Washington Post* (23 July 1981) and *Chicago Tribune* (21 August 1981), and an article by science writer Isaac Asimov was published in the *New York Times Sunday Magazine* (14 June 1981).

Extensive coverage of the Arkansas trial often was linked to a local story. The interest of the *New Orleans Times-Picayune* was clearly pegged to the American Civil Liberties Union challenge of a similar Louisiana law; coverage in the *St. Louis Post-Dispatch*, to a pending Supreme Court ruling on religious activity at the University of Missouri. The *Houston Chronicle* referred to the Arkansas decision as the "flip side" of a Houston court case in the early 1970s (6 January 1982).

During the trial, local columnists and editorial cartoonists at many major papers found grist for the mills of satire; and syndicated columnists such as George Will and Ellen Goodman also commented on the significance of the case. Editorials on the trial and Judge Overton's decision appeared in, for example, the *Lincoln* (NB) *Journal Star, Chicago Tribune*, and *Detroit News* and in many Southern newspapers.

In addition to general coverage of fundamentalists, both *Time* and *Newsweek* carried several reports on the Arkansas case, in March and December 1981 and January 1982. The newsmagazines treated it, however, not as science but as religion, law, or education. Scientists were quoted only rarely—the *Newsweek* report of 21 December 1981, for example, mentions Darwin, Copernicus, Galileo, and six attorneys, but only one contemporary scientist (Francisco Ayala, who had just testified in Arkansas). Accounts stressed the difficulties that court and counsel encountered in grappling with highly technical testimony; yet none of the scientific testimony was described.

Even though they often ignored the science, general-audience magazines did not refrain from criticism or comment on the scientists. Both *The New Republic* (4 April 1981) and *The Nation* (21 March 1981) published vigorous critiques of the scientific community's reaction to the creationists, warning of possible "back-fire"; articles praising the scientists' actions in forming "Committees of Correspondence" to "combat" the creationists' efforts to influence local school boards appeared in other partisan publications such as *The Humanist*. Some publications made exceptional efforts to give the creationists space for debate; see, for example, the 19 March 1982 issue of *Christianity Today*. Following the decision, articles on the case and its effects appeared in magazines ranging from *Harper's* to *Mademoiselle*, as well as in the popular science journals such as *Discovery, Science 81*, and *Science News*; some of the most balanced reports appeared in publications directed at the business community, such as *Publishers Weekly*.

There is no evidence that newspapers or newsmagazines made any special effort to assign reporters who had any expertise in science journalism, even though substantial portions of each day's testimony contained complex scientific evidence for and against evolution and discussion of the philosophy of science and the scientific research system. Only two major newspapers—the *Boston Globe* and *Washington Post*—carried articles on the trial written by their science writers. (The *New York Times* did run a related article—primarily on evolutionary theory—by Walter Sullivan during the course of the trial.) It was more common for a paper to assign a political or legal affairs reporter to the story.

This downplaying of science is further evident in the way the witnesses were covered. Of the four research scientists who testified for the plaintiffs (Francisco Ayala, Brent Dalrymple, Harold Morowitz, and Stephen Jay Gould), none were quoted or even mentioned in the trial coverage in the *New York Times, Chicago Tribune, Baltimore Sun, Detroit Free Press, Kansas City Star, Los Angeles Times*, and many other papers; only two (Ayala and Dalrymple) were mentioned in the *Time* and *Newsweek* stories; some papers, such as the *Washington Post*, reported on the testimony of just one witness. Only two major papers outside Arkansas (papers in New Orleans and St. Louis) mentioned all four of the plaintiffs' science witnesses. Yet the *New York Times* and the *Detroit News*, for example, reported on the testimony of six out of eleven witnesses for the state. One witness upon whose research and writings many of the news summaries of the history of Creationism undoubtedly relied and whose work and testimony was cited repeatedly by Judge Overton in his Opinion—sociologist Dorothy Nelkin—was mentioned in the news reports in one of eighteen major newspapers reviewed for this study—and then only in a "wrap-up" story on the trial, not on her individual testimony.[13]

Few scientists (and none who had not testified in Arkansas) were quoted or mentioned in the news accounts, and most news stories did not adequately explain the scientific testimony, despite the amount of scientific evidence given and even though the weight of the state's argument for the constitutionality of Act 590 rested on the attempt to prove creation-science a legitimate "alternative" theory of origins. Journalists did write about testimony on the "accepted standards for consideration as a science," but far more emphasis was placed on the social and political aspects and on the trial as a symbol of the persistence of fundamentalist, politically conservative principles in the "New South."[14] For example, only on 27 December 1981 did the *New York Times* run an article that connected the scientific debate over evolution and the origin of the universe to the theories of creation-science.[15]

Nor was there clear exposition of the legal precedents. The Scopes trial was presented in its social context and the distinction between the two cases was rarely made clear. (In fact, if the readers learned anything, we may asssume it is the details of the Scopes trial. Most stories on the Arkansas trial mentioned Scopes; several papers, including the *New York Times*, used identical capsule paragraphs on Scopes in *all* of the early coverage of the trial. Newspapers and magazines ran photos of Darrow and Bryan at Dayton; *Newsweek* called the Arkansas case just a "mutant" of Scopes. John Scopes, in fact, was mentioned far more than any of the Arkansas witnesses.) Instead, the articles emphasized the creationists' political arguments for "equal time" and were classified as something other than science.

The news coverage, then, mostly depicted the creationists as a beleaguered but sincere minority only seeking fair play for their somewhat untraditional scientific ideas. When scientists were quoted in the news, they were often described as "angry" or as "scornful" of creationists' arguments, their statements were critical or full of hostile or loaded words, and they frequently accused the creationists of deception. The scientists' theme was not, however, echoed in the newspaper editorials, which tended to describe creationists as misdirected (and possibly dangerous) religious zealots and which were not explicitly supportive of the "main-line" scientific community.

Local Coverage of the Trial

Two newspapers dominate journalism in Arkansas and have done so for well over a hundred years. The *Arkansas Gazette*, founded in 1819, bills itself as "The Oldest Newspaper West of the Mississippi." It won Pulitzer prizes for reporting on the Little Rock school integration crisis in the 1950s and is regarded as a liberal paper, although not linked consistently to a particular political party. The *Arkansas Democrat*, founded in 1871, is a politically conservative afternoon paper that has been supportive of the Democratic party in Arkansas. Both are serious newspapers in the old tradition and both stress reporting on state and local government above international news or tabloid sensationalism.

To understand the tone of the Arkansas news reports it is necessary first to understand their context. Arkansas is a state which annually spends tax money for a nativity scene on the front lawn of the state capital at Christmastime (*Gazette*, 13 December 1981), in which the Little Rock School Superintendent approved two conservative publications for classroom use, one of them published by the John Birch Society (*Democrat*, 16 December 1981), and in which a member of the Ku Klux Klan openly demonstrated outside the courthouse where the *McLean* case was being tried (*Democrat*, 18 December 1981). Religion is an accepted part of daily life, even for liberals, and it is within that context that such headlines as "Creation Took Week in Bible, Trial Expected to Last Two" (*Gazette*, 6 December 1981) must be read.

Both papers covered the trial well. The principal difference between them occurs in how each defined the trial's focus. The *Gazette* reported that the ACLU had raised three issues in challenging the constitutionality of Act 590: whether the Act violated the Establishment Clause of the First Amendment; whether it violated rights of academic freedom; and "whether the law is too vague for the ordinary person to understand what is required or prohibited by it" (*Gazette*, 6 December 1981). In a companion story, the *Gazette* labeled the "key" issue as whether it is possible to separate creation-science from its religious base because the bill "specifically says that no religious materials or instruction can be used in presenting creation-science" (*Gazette*, 6 December 1981). The *Democrat* seemed to assume that the trial would debate the "truth" of evolution. Despite the fact that the Arkansas Attorney General had declared only two days before that "God is not on trial next week . . . we will not prove that God created man or that man descended from apes,"[16] the *Democrat* pronounced that "on trial will be more than the issue of what theories of man's origins should be taught. . . . Also at issue . . . will be the constitutional separation of church and state, the state's right to mandate public school curriculums, and the ability of special interest groups to influence the laws of the land" (*Democrat*, 6 December 1981).

Both papers carried photographs of the empty courtroom and printed witness lists, imparting a sense of the drama to come. On the Sunday before the trial opened, the *Democrat* printed a special three-page section that included the text of Act 590, photographs of all the attorneys in the case, a photograph and story on Carl Sagan's withdrawal as a witness, a history of the Scopes trial, analyses of the potential effect of the Act, and a cartoon drawing of a human hand (white) reaching down to touch fingertips with a simian hand (black). A *Democrat* political columnist predicted that the trial would be "a three-ring circus."

On Tuesday, December 8, following the opening of the trial, a *Democrat* news story complained that the judge had "allowed" the ACLU to "introduce religion" into this "land-mark origins trial," and an editorial identified the Act's intention as the "free flow of ideas in the classroom" and asked, "Who'd quarrel with that?" The implication was that the ACLU had a "crabbed" view of individual rights. The *Gazette* described the same first-day testimony as "Creation Science Law Religious in Nature, Theologian Testifies" (8 December 1981).

In coverage of the second day of the trial, the *Democrat* reported that "ACLU witnesses testify creationism not science," but the story emphasized Michael Ruse's testimony regarding the "tentative" nature of science, that is, that science is "not necessarily the final word" (9 December 1981). The *Gazette* for that day crowed, "Creationism Premises False, Twist Science, Trial on Act 590 Told" and reported on Ruse's analysis of creation-science rather than on his definition of science.

Such differences continued throughout the trial. *Gazette* editorials called for the testimony to "stick to the central issues swirling about the Constitution's establishment clause" (9 December 1981). In coverage of the third day, the *Democrat* emphasized the testimony of educators regarding the effects in the classroom; the *Gazette* continued to emphasize that creation-science is bound to religion and is not science under any conventional definition, repeating this theme in subsequent stories as the state began to present its witnesses (e.g., "Genesis Basis for Law, Two Witnesses Agree," 12 December 1981; "Religious Dimensions of 'Creation-Science,' " 16 December 1981). The *Democrat* argued that to find Act 590 unconstitutional would be "a perversion of academic freedom." "Establishmentarianism science rules," a *Democrat* editorial complained, and therefore creation-science has no chance for "fair play" in the classroom. The newspaper's managing editor declared that "pushy evolutionists curtail academic freedom" (29 December 1981), and another *Democrat* story carefully pointed out that, although the popular image of the ACLU was as a "champion of unpopular causes," in this case the organization was defending "conventional" scientific beliefs and the status quo:

The ACLU called to the witness stand what it considered to be the nation's top scientific minds and asked them to climb down from their ivory towers and say that creation science, in their opinion, was not legitimate.[17]

When on January 6, the *Democrat* announced "Judge voids creationism law," its lead editorial ("Hook, line, and sinker") complained that Judge Overton could have "said so much positively about the claims of academic freedom and free speech," but that instead he had dwelt only on "the invalidity of creation science." He had, according to the *Democrat*, "bought" the ACLU's "barren negativistic arguments of sneaking religion" and the state's only recourse was to appeal immediately and to challenge this "unenterprising handling of a great issue." The *Gazette* declared that the ruling reaffirmed the constitutional principle of the "wall of separation between church and state," but pointed out that, in the long run, one of the Act's original purposes may have been accomplished. In evidence, the paper quoted the reaction of James L. Holsted, the State Senator who had introduced Act 590:

When you consider what I was trying to do, we've been victorious. I feel like we really won because people are talking about it, kids will be asking about it. Teachers will have to talk about creation-science. . . . In fact, it's just starting. All the hoopla and publicity—that's just what I wanted. . . . I wanted it to get the attention and make people talk about it.[18]

The *Gazette* also advocated the prompt appeal of the ruling—not with the intention of overturning it, but to reaffirm the principle of separation of church and state. Nevertheless, the day after the state finally announced it would not appeal, *Gazette* editorial cartoonist George Fisher showed Act 590 being buried in "Good Riddance Cemetery" (7 February 1982).

Although each paper assigned veteran political reporters to cover the trial, neither provided special science coverage, and very little of the scientific or technical testimony was reported in local papers. It may be that, in addition to a feeling of lack of expertise in interpreting scientific information, the Arkansas newspapers simply regarded the actual scientific testimony as the least important aspect of the trial, containing information of interest to scientists but not germane to the central questions of fair play, academic freedom, and establishment of religion.

Two social aspects of science were covered in the Arkansas papers: the standards for what is or is not a science, and the response of the scientific community to creation-science—but, again, not fully. Although both newspapers frequently mentioned the debate over the "standards" for what is science, especially in connection with the testimony of Michael Ruse and Francisco Ayala, neither paper reported, for example, on Ruse's list of criteria or explained some of the criteria mentioned throughout the trial. The attitude toward science was therefore ambiguous. The conviction that "no amount of theorizing on origins is going to be conclusive" (*Democrat*, 8 December 1981) acknowledged that science is mutable and *open* to change. Still, the editorial writers seemed to seek assurance, to need some intellectual "Rock of Ages" as a measure of scale. "[D]ifferences of scientific opinion on such topics as thermodynamics and radiometrics . . . remain open to argument and proof, subject to scientific testing, not judicial fiat," the *Democrat* declared (10 January 1981).

Such discussions convey an image of science as the autocratic arbiter of its own rules. One scientific witness for the plaintiffs stated: "Science is what is accepted in the scientific community. The community has rules by which it operates" (*Democrat*, 10 December 1981)—a statement that surely must appear dogmatic and "closedminded" to the nonscientist. This image was further reinforced by the Attorney General's outline of what he called his "country club" defense:

. . . in an effort to develop the thesis that creationists are shut out from an elitist scientific establishment partly out of fear of the threat of a new idea . . . [the Arkansas Attorney General] said his staff would attempt to show that established scientists were trying to "protect and enrich their positions in the scientific community."[19]

and by the creationists' complaints of censorship:

"New or radical ideas that do not conveniently fit into world views of scientists are frequently not given the slightest consideration," [Morrow] said.[20]

The Scientific Press

Coverage of the Arkansas trial and Creationism in the news and review magazines published for the scientific community offers some further indication of attitudes within the scientific community which may contribute to public sympathy for the creationists.

Concern about the effect of the creationists' efforts extends beyond the field of biology. In *Physics Today*, for example, we find that "during the last few months it has become apparent to many physicists that the movement that calls itself creation-science is addressing scientific matters outside biology and is threatening to transform science education in this country."[21] The primary impetus for the physicists' concern is the potential effect of similar laws on the teaching of physics; *Physics Today* also mentioned that the case "pointed to the power of the creationists and gave them the legitimacy of advocating one side of a controversial issue," and the magazine admonished that "it is no longer appropriate to withhold a reaction to the creationists in order to deny their arguments the dignity of a scientific refutation."

Shortly after the *McLean* decision, *Chemical & Engineering News* published a comprehensive report on creationists' efforts in Arkansas in relation to the whole of American science, chemistry as well as evolutionary biology. In an editorial ("Creationism: It matters a lot") published in the same issue, the editor of *Chemical & Engineering News* reminded his audience that the *McLean* case "is but one manifestation of the antiscience attitude that pervades much of society today."[22] The editorial described the "threat" that Creationism poses to science and the need to "rebuff such attacks" and "assaults."

The most substantial reporting on the creationists done by a science journal has appeared in the "News and Comment" section of *Science*, the official publication of the American Association for the Advancement of Science. Creationism was an issue in its pages long before Act 590. *Science* was, for example, one of the few journals to discuss the implications of the statement by (then presidential nominee) Ronald Reagan supporting the creationists' right to equal time in the public schools. To *Science*, such high-level support raised the question of the "extent to which creationists may be able to affect school curriculums in the future."[23] A subsequent *Science* article referred to a creationist suit in California as "the prelude to the rumble of heavy artillery."[24] When the Arkansas trial drew closer, *Science* reported on coordinated efforts to "respond" to the creationists and again used the language of confrontation ("the threat of creationism," "science is under attack," "the fight will be on many fronts," "immediate skirmish"). The articles on the trial itself had an unfortunately pugilistic tone, frequently ridiculing the clothing and appearance of the creationists or their supporters. A somewhat more considered analysis of the case appeared later in the trial series; yet even in that report derogatory remarks were allowed to stand in print—for example, the state's creation-science witnesses were referred to as "presumably the best of the best" and described as having "burdened" themselves with admissions on the stand.[25]

Television Coverage

William J. Broad, writing of the California textbook suit brought by creationists, observed that "Possibly their most significant gain was attention from the press. The media blitz alerted the wider fundamentalist community to the evolution issue in general."[26] Television offers the quickest route for creationists to send out their messages and, as are many of the modern American fundamentalists, the creation-scientists are skilled performers in electronic evangelism.

During the trial, the principal attorneys and some of the witnesses were interviewed on such programs as "NBC Magazine" (Saturday, 5 December 1981), "Good Morning, America" (ABC, 7 December 1981 and 6 January 1982), and local news interview programs. (The Arkansas Attorney General was also the object of a brief skit on "Saturday Night Live" (NBC) on 12 December 1981.) But, because of the court's restrictions on filming testimony, television coverage of the trial was limited to interviews in the courthouse corridors or in studios.

More considered discussion took place in two documentaries shown on public television stations throughout the country—one in the WGBH-TV "Nova" series, the other in the KPBS-TV "Synthesis" series. A comparison of the tone and production of the two programs highlights some of the difficulties in unbiased presentation of the creationists' views.

The "Nova" broadcast, "Did Darwin Get It Wrong?" was originally produced by the British Broadcasting Corporation and then purchased for WGBH-TV, which added an "update" on the Arkansas case at the end of the film.[27] It was first broadcast in the U.S. a few weeks before the *McLean* trial began, and it adheres to the typical format of the "Nova" series: glorious photography, complex script, and soporific tone.

Most of the visual presentations in the program are films either of birds, animals, or the faces of talking scientists, interspersed with film of laboratory equipment.

"Nova" publicity described the program as showing that "the evidence for evolution is not nearly as solid as most people assume." Nevertheless, although the film begins and ends with statements by or about creationists and their beliefs, it is in fact predominantly about competing explanations of evolution, and confuses the creationists' critique and rejection of all evolutionary theory with the scientific debate over just how evolution took place.

The narration is unashamedly authoritative, and all the scientists interviewed have a curious coincidence of costume—the evolutionists are informally dressed (open-collar shirts and sweaters) while all the creationists or critics of evolution are filmed in business suits and formal setting. Such distinctions are useful to the viewer, however, because there is inconsistent, incomplete, and occasionally nonexistent identification of speakers—one person is identified only by name and birthplace, another only as "a lawyer."

After reviewing the "Nova" program in detail, it is not hard to see why a creationist writing in a letter to the *Arkansas Democrat* described the program as voicing "very real scientific disenchantment with Darwinism" and held it as proof for his own beliefs. The program gives this impression even though statements by only one creationist are included.

Another public television documentary on Creationism, produced for the KPBS-TV "Synthesis" series on science and public policy, shows creationists on camera for over one-quarter of the total film (approximately 15 minutes out of 59), yet it overwhelmingly builds a case *against* admitting creation-science to the science curriculum. The program, "Creation vs. Evolution: Battle in the Classroom,"[28] uses interviews with parents and children caught up in a local textbook controversy to set the stage not (as in "Nova") for a confrontation between science and the creationists, but for a "battle for the minds of children."

More than twenty minutes elapse in the "Synthesis" film before any scientist (creationist *or* evolutionist) is interviewed. The opening "debate" is instead constructed about interviews with Rev. Tim LaHaye, founder of the Institute for Creation Research, and Rev. Bill Nebo, a Presbyterian minister who believes in evolution. The arguments made by LaHaye and Nebo are simple, concise, and articulate versions of the same feelings expressed in, for example, letters to the Arkansas newspapers during the *McLean* trial. LaHaye talks about changing moral values and equates evolution with social deterioration. Nebo explains how many people "want to have life hold together in some kind of simple coherent system," and asserts that Fundamentalism and Creationism offer that. To LaHaye, "design demands a designer"; to Nebo, the complexity of evolution is a manifestation of the glory of God. The narrator carefully points out that God, or supernatural causes, is considered out of the purview of the scientific mind and that Creationism's acceptance of supernatural causes makes it unacceptable to scientists.

The film then focuses on the debate between scientists and creation-scientists over particular theoretical points—for example, the age of the earth and how it is measured, or the Noachian Deluge. Like the "Nova" film, evolutionists and creationists are distinguishable by the formality of their clothes. Unlike the "Nova" film, speakers are clearly identified by title and affiliation as well as scientific field.

The scientists do not refrain from accusation or the use of hostile language on camera; one refers to "typical creationist baloney," "so-called creation-scientists," and creationist "hokum" and "gibberish," while the creationist spokesman gently pleads that "our science is just as good as anyone else's." Yet the creationists' evangelistic intentions are made explicit through their own words, in interviews, and through samples of creationist films (e.g., "Footprints in Stone," a program produced by Films for Christ).

The central creationist theme simply of wanting equal time for their beliefs is given full exhibit in an interview with Tim LaHaye as he explains why the scientists protest so much about creationists:

... the evolutionists have had it all their way for seventy-five years—particularly since John Dewey. Now the fact is that they will have to let us in or let someone come in with an alternate view to theirs. They don't like it, but they'll have to live with it because the overwhelming majority of the American people believe in creation. . . .

A few minutes later, that theme is rephrased as Nell Segraves of the Creation-Science Research Center accuses scientists of acting like "high priests . . . in a sacred position." By arguing that the "two models" (creation-science and evolutionary theory) can live side by side, the creationists thus issue a direct challenge to sacred authority.

Through the use of film clips from the creationists' own films and by showing the movement's political links to the Moral Majority, the "Synthesis" documentary makes clear the creationists' evangelistic motives without ridiculing (or, indeed, questioning) their right to maintain such beliefs or to build private institutions to research their beliefs, even in a scientific country. The real danger, the film makes clear in its closing scenes, comes from the attempt of a religious group to impose its beliefs on the public schools by masquerading those beliefs as science.

Some Questions of Value and Conspiracy

Herbert Gans, in *Deciding What's News*, maintains that "there is a difference between the values in the news and the value implications of the news . . . while the former may ultimately originate with the journalists, the latter do not."[29] The value implications of news coverage of the Arkansas case are two-fold: (a) in providing publicity for the creationist movement and (b) in either confirming or questioning the place of science in modern American society.

From the standpoint of Gans' analytical scheme, the news reporting on the Arkansas case may be regarded as having stressed two values in particular. One is *moderatism*, an attitude Gans explains thus: "individualism which violates the law . . . is suspect; equally important, what is valued in individuals is discouraged in groups . . . [and] polar opposites are questioned. . . ."[30] In the news coverage, the creationists are called into question not because of their specific beliefs (which may be shared by many other Americans) but because they have, as a group, attempted to impose their beliefs on others, through the public schools, in violation of the Constitution. Moreover, creationists appear to represent a religious extreme because their beliefs about evolution are not shared by the "main-line" churches, even in the South. Most of the news reports stressed the interdenominational nature of the challenge to Act 590 (the plaintiffs included rabbis and Catholic priests as well as Protestant ministers), and most covered the testimony of Methodist Bishop Kenneth Hicks, one of the plaintiffs, though they did not cover the testimony of equally critical science or education witnesses.

The other value expressed in the news reports was that of *moral order*. Although the creationist movement itself has been characterized as a reaction to the perception of declining national mores, the strong political activities of the creationists (and their overt links to the Moral Majority) evoke an image of a threat to moral (and possibly to political) order. The attempt "to return to basic values" espoused by the creationists may be perceived by some in the media as also embracing a retrenchment of the Progressive ideal.[31] Moreover, the creationists derive their political strength from the lower middle class (something stressed in many of the "elite" publications such as the *New York Times* and

Harper's), and are attempting to do battle with the scientific community, an acknowledged institution of the elite.[32] The image of a battle between unequal forces is heightened by the hostile public statements of scientists.

The creationists play on this theme again and again in their interviews. Attorney Richard Turner, who fought the California textbook challenge in March 1981, characterized the scientists thus:

They get up on the stand and act as if their very lives were being attacked. They not only close ranks, but they almost deny anybody the right to know of the internal fights that go on within the evolutionary crowd. They're pompous and arrogant, just the kind of people that the First Amendment was written to protect us against.[33]

Both groups cry "conspiracy" in the press. For example, an article in the 29 January 1982 issue of *Science* investigated the possibility that Arkansas Act 590 had been "the result of a conservative 'conspiracy.' "[34] The article described more about down-home Southern politics than about science, but it left hanging the central question—conspiracy against whom or what? Against science? The creationists, of course, would like the public to believe there is a conspiracy of scientists who are aiming to destroy religion and "propagate immorality" through the teaching of evolution.[35] An 8 January 1982 editorial in the *Arkansas Democrat* referred to scientists who were beginning to take political action against the creationists as "evolutionist street fighters," as "dishonest science zealots" who were part of a "conspiracy." That editorial ominously advised scientists to limit their activities "to the adducement of evidence and the writing of scholarly articles on specialties."

Science journalist Philip J. Hilts has observed that the Arkansas trial did "what no other creation vs. evolution trial, including the famous Scopes 'monkey trial,' has done: allowed a head-on collision between religion and science."[36] This review of the news coverage of the Arkansas trial indicates that the real confrontation is once again taking place not in the churches or even in the classrooms but in the public arena dominated by the mass media.

Acknowledgment

I wish to thank Henry L. Chotkowski and David Klasfeld for their help in obtaining copies of the newspaper stories on the Arkansas trial.

Notes and References

1. Quoted in "The Arkansas Press" section, *Arkansas Gazette* (6 December 1981); also see Meredith Oakley, "Sorry, Steve, it looks like Scopes II," *Arkansas Democrat* (6 December 1981).

2. Michael Williams, "At Dayton, Tennessee," *Commonweal* (22 July 1925): 262.

3. John T. Scopes and James Presley. *Center of the Storm: Memoirs of John T. Scopes* (New York: Holt, Rinehart, and Winston, 1967), p. 183.

4. Eric Barnouw, *A Tower in Babel: A History of Broadcasting in the United States to 1933* (New York: Oxford University Press, 1966), p. 196. Barnouw comments that the media publicity over Scopes contributed to the introduction in the 69th Congress of an amendment to ban all "discourse" on evolution from the radio. The measure was defeated. (p. 197)

5. Scopes, *op. cit.*, p. 102.

6. Williams, *op. cit.*, p. 262.

7. See, for example, Reginald Stuart, "Judge's Conduct of Trial is Praised," *New York Times* (17 December 1981): A27; and Rone Tempest, " 'Creationism' Judge Avoided Air of Circus," *Los Angeles Times* (19 December 1981).

8. For example, Judge Perluss, in restricting the number of reporters who were allowed seats at the California textbook guidelines hearing, stated "I don't want this to be a media performance" (Wallace Turner, "New Trial on Evolution Off to Slow Start on Coast," *New York Times*, 3 March 1981: A14).

9. An *Arkansas Democrat* reporter complained after the trial ended that the witnesses "at times, . . . ignored the judge and talked directly to reporters, many of whom were seated in the jury box" (Larry Ault, "Roles switched in Act 590 trial," *Arkansas Democrat*, 20 December 1981: 23A).

10. "Seventy-five News Organizations Register to Cover Trial," *Arkansas Gazette* (20 December 1981).

11. Leonard Granato, "Creation-Science and the Outside Press," *Arkansas Gazette* (19 March 1982): 15A.

12. The conclusions in this section are based on a review of news reports and articles in several dozen newspapers and magazines, listed in the New York Times Information Bank, and Bell & Howell Newspaper Index (covering ten papers), Bell & Howell Index to Black Newspapers (covering eleven papers), and Reader's Guide for general periodicals for the period 1 March 1981 through 1 February 1982 (for newspapers) and through July 1982 (for periodicals). According to the Bell & Howell Index search, no substantive news articles or editorial columns on the Arkansas case appeared in the eleven black-oriented newspapers covered by that Index.

13. Daniel McShea, "The nature of science is on trial in Scopes II case," *Boston Globe* (14 December 1981).

14. See, for example, Steven M. Luxenberg, " 'To tell the old, old story': Pews are full in Arkansas as state tests creationist law," *Baltimore Sun* (13 December 1981): A1, A19; Roy Reed, "In Rural Arkansas, Evolution Betokens Baleful Worldliness," *New York Times* (18 December 1981).

15. Walter Sullivan, "Creation Debate Is Not Limited to Arkansas Trial," *New York Times* (27 December 1981): A48.

16. "Exciting to Try Suit, Clark Says; Stresses God Not on Trial," *Arkansas Gazette* (5 December 1981): 5A.

17. Larry Ault, "Roles switched in Act 590 trial," *Arkansas Democrat* (20 December 1981): A1.

18. "Sponsor Sees Victory In End on Act," *Arkansas Gazette* (6 January 1982): A1.

19. George Wells, "Creationism Is Bound to Religion, Educators Say in 4th Day of Trial," *Arkansas Gazette* (11 December 1981): 1, 9.

20. Defense witness W. Scott Morrow, quoted in "Censorship Charge Not Supported, Judge Complains," *Arkansas Gazette* (15 December 1981): 1A, 7A.

21. "Mainstream Scientists Respond to Creationists," 35 *Physics Today* 2 (February 1982): 53–55.

22. Michael Heylin, "Creationism: It Matters a Lot," p. 7, and Rudy M. Baum, "Science Confronts Creationist Assault," pp. 11–26, 60 *Chemical & Engineering News* 3 (18 January 1982).

23. "Republican Candidate Picks Fight With Darwin," 209 *Science* (12 September 1980): 1214.

24. William J. Broad, "Creationists Limit Scope of Evolution Case," 211 *Science* (20 March 1981): 1331–1332.

25. Reports by Roger Lewin on the Arkansas Creationism case appeared in 214 *Science* (6 November 1981; 4 December 1981; 11 December 1981) and 215 *Science* (1 January 1982; 8 January 1982).

26. Broad, *op. cit.*, p. 1332.

27. "Did Darwin Get It Wrong?" "Nova" series, WGBH-TV, Boston; written, produced, and directed by Alec Nisbett; first broadcast on PBS in November 1981.

28. "Creation vs. Evolution—Battle in the Classroom," "Synthesis" series, KPBS-TV, San Diego; produced and directed by Ryall Wilson, written by Anne Mendelsohn; first broadcast Spring 1982.

29. Herbert Gans, *Deciding What's News* (New York: Pantheon, 1979), p. 40.

30. *Ibid.*, p. 51.

31. *Ibid.*, p. 69.

32. See *ibid.*, p. 61, for a discussion of elites in the news.

33. Quoted in Broad, *op. cit.*, p. 1332.

34. Roger Lewin, "A Tale with Many Connections," 215 *Science* (29 January 1982): 484–487.

35. H. L. Mencken wrote of Scopes' accusers that they believed "there is . . . a conspiracy of scientists afoot" whose purpose is "to break down religion, propagate immorality, and so reduce mankind to the level of a brute." Quoted in the *New York Times* (12 January 1982): A15.

36. Philip J. Hilts, "Religion Influenced Ark. Legislator Who Wrote Creationism Law," *Washington Post* (14 December 1981).

Select Bibliography*

I. Science and Religion

A. General Discussion

Barbour, Ian G. *Issues in Science and Religion* (Englewood Cliffs, NJ: Prentice-Hall, 1966).

Barbour, Ian G. *Science and Religion: New Perspectives on the Dialogue* (New York: Harper and Row, 1968).

Booth, Edwin Prince. *Religion Ponders Science* (New York: Appleton-Century-Crofts, 1964).

Gilkey, Langdon. *Religion and the Scientific Future: Reflections on Myth, Science, and Theology* (Macon, GA: Mercer University Press, 1970).

Habgood, J. "The Uneasy Truce Between Science and Religion," in A. R. Vidler, ed., *Soundings* (Cambridge, England: Cambridge University Press, 1963), pp. 21–41.

Jaki, Stanley L. *The Road of Science and the Ways of God* (Chicago: University of Chicago Press, 1978).

Shapley, Harlow. *Science Ponders Religion* (New York: Appleton-Century-Crofts, 1960).

White, Edward A. *Science and Religion in American Thought: The Impact of Naturalism* (New York: AMS Press, 1952).

B. Evolution, Creation, and Religion

Benz, Ernst. *Evolution and Christian Hope: Man's Concept of the Future from the Early Fathers to Tielhard de Chardin*, Heinz G. Frank, trans. (Garden City, NY: Doubleday, 1966).

*Compiled by Melinda Thomas, Massachusetts Institute of Technology.

de Mille, Richard. "And God Created Evolution." 34 *National Review* 5 (19 March 1982): 288, 292. (Sorts through the differences between science and religion and suggests reasons why there is a Creationism controversy.)

[Eggleston], Sister Mary Frederick. *Religion and Evolution Since 1859: Some Effects of the Theory of Evolution on the Philosophy of Religion* (Chicago: Loyola University Press, 1935).

Ewing, J. Franklin, S. J. "Current Roman Catholic Thought on Evolution," in Sol Tax and Charles Callender, *Evolution After Darwin*, vol. 3 (*Issues in Evolution*) (Chicago: University of Chicago Press, 1960), pp. 19–28.

Fitch, Robert Eliot. "Darwinism and Christianity." 19 *Antioch Review* (Spring 1959), pp. 20–32.

Foster, Michael B. "The Christian Doctrine of Creation and the Rise of Modern Natural Science." 43 *Mind* (October 1934): 446–468.

Gilkey, Langdon. *Maker of Heaven and Earth: A Study of the Christian Doctrine of Creation* (Garden City, NY: Doubleday, 1959).

Hayes, Zachary, O. F. M. *What Are They Saying About Creation?* (New York: Paulist Press, 1980).

Henry, Carl. "Theology and Evolution," in R. Mixter, ed., *Evolution and Christian Thought Today* (Grand Rapids: Eerdmans, 1959), p. 218.

Jaki, Stanley L. *Creation and Science: From Eternal Cycles to an Oscillating Universe* (Edinburgh: Scottish Academic Press, 1977).

Lack, David. *Evolutionary Theory and Christian Belief: The Unresolved Conflict* (London: Methuen, 1961).

Pannenburg, Wolfhart. "Theological Questions to Scientists." 16 *Zygon* 1 (March 1981): 65–77.

Peacocke, A. R. *Creation and the World of Science: The Bampton Lectures, 1978* (Oxford: Clarendon, 1979).

Peacocke, A. R., ed. *The Sciences and Theology in the Twentieth Century* (Notre Dame: University of Notre Dame Press, 1981).

Pelikan, Jaroslav. "Creation and Causality in the History of Christian Thought," in Sol Tax and Charles Callendar, *Evolution After Darwin*, vol. 3 (*Issues in Evolution*) (Chicago: University of Chicago Press, 1960), pp. 29–40.

Schwarz, Hans. "The Significance of Evolutionary Thought for American Protestant Theology: Late Nineteenth-Century Resolutions and Twentieth-Century Problems." 16 *Zygon* 3 (September 1981): 261–284.

Wysong, R. L. *The Creation-Evolution Controversy* (East Lansing, MI: Inquiry Press, 1976).

II. Fundamentalism

Barr, James. *Fundamentalism* (Philadelphia: The Westminister Press, 1978).

Cole, Stewart G. *The History of Fundamentalism*, reprint edition (Westport, CT: Greenwood Press, 1971 [1931]).

Dollar, George W. *A History of Fundamentalism in America* (Greenville, SC: Bob Jones University Press, 1973).

Dollar, George W. "The Early Days of American Fundamentalism." 123 *Bibliotheca Sacra* (1966): 115–123.

Furniss, N. F. *The Fundamentalist Controversy, 1918–1931* (New Haven: Yale University Press, 1954).

Marsden, George M. *Fundamentalism and the American Culture: The Shaping of Twentieth Century Evangelicalism, 1870–1925* (New York: Oxford University Press, 1981).

Peter, Walter G., III. "Fundamentalist Scientists Oppose Darwinian Evolution." 20 *Bioscience* 19 (1 October 1970): 1067–1069.

Sandeen, Ernest R. "Toward a Historical Interpretation of the Origins of Fundamentalism." 36 *Church History* (March 1967): 66–83.

Sandeen, Ernest R. *Roots of Fundamentalism: British and American Millenarianism, 1800–1930* (Grand Rapids: Baker Book House, 1978).

III. Darwinism

A. Some Nineteenth-Century Responses
Aveling, Edward B. *The Religious Views of Charles Darwin* (London: Free Thought Publishing Company, 1883).

Bascom, John. *Evolution and Religion; or Faith as Part of a Complete Cosmic System* (New York: G. P. Putnam's Sons, 1897).

Curtis, George Ticknor. *Creation or Evolution? A Philosophical Inquiry* (New York: D. Appleton & Company, 1887).

McCosh, James. *The Religious Aspect of Evolution* (New York: Scribner's, 1890).

B. Historical Analyses of the Response to Darwinism
Appleman, Philip, ed. *Darwin* (New York: Norton, 1970).

Betts, John Rickards. "Darwinism, Evolution, and American Catholic Thought, 1860–1900." 45 *Catholic Historical Review* 2 (July 1959): 161–185.

Dupree, A. Hunter. "The First Darwinian Debate in America: Gray Versus Agassiz." 88 *Daedalus* (Summer 1959): 560–569.

Ebenstein, William. "The Early Reception of the Doctrine of Evolution in the United States." 4 *Annals of Science* 2 (15 April 1939): 306–321.

Farley, John. "The Initial Reactions of French Biologists to Darwin's 'Origin of Species.' " *Journal of the History of Biology* 7 (1974): 275–300.

Gillispie, Charles C. *Genesis and Geology* (Cambridge, MA: Harvard University Press, 1951).

Glick, Thomas F., ed. *The Comparative Reception of Darwinism* (Austin: University of Texas Press, 1974).

Gould, Stephen Jay. "Agassiz in the Galapagos." *Natural History* (December 1981): 7–14. (Postulates that Agassiz visited the Galapagos to "test" the theory of evolution and ended up "defending" Creationism.)

Gould, Stephen Jay. *Ever Since Darwin: Reflections in Natural History* (New York: Norton, 1977).

The Impact of Darwinian Thought on American Life and Culture: Papers Read at the Fourth Annual Meeting of the American Studies Association of Texas at Houston, Texas (Austin: University of Texas Press, 1959).

Kelly, Alfred. *The Descent of Darwin: The Popularization of Darwinism in Germany, 1860–1914* (Chapel Hill: University of North Carolina Press, 1981).

Loewenberg, Bert James. "Darwinism Comes to America, 1859–1900." 28 *Mississippi Valley Historical Review* (December 1941): 339–368.

Moore, James R. *The Post-Darwinian Controversies: A Study of the Protestant Struggle to Come to terms with Darwin in Great Britain and America, 1870–1900* (Cambridge, England: Cambridge University Press, 1979).

Persons, Stow, ed. *Evolutionary Thought in America* (New Haven: Yale University Press, 1950).

Roberts, Winsor H. "The Reaction of American Protestant Churches to the Darwinian Philosophy, 1860–1900." Ph.D. dissertation, University of Chicago, 1936.

Ruse, Michael. *The Darwinian Revolution: Science Red in Tooth and Claw* (Chicago: University of Chicago Press, 1979).

IV. The Scopes Trial

Allen, Leslie H., ed. *Bryan and Darrow at Dayton: The Record and Documents of the "Bible-Evolution" Trial* (New York: A. Lee & Company, 1925).

de Camp, L. Sprague. *The Great Monkey Trial* (Garden City, NY: Doubleday, 1968).

Gatewood, W. B., Jr., ed. *Controversy in the Twenties. Fundamentalism, Modernism, and Evolution* (Nashville: Vanderbilt University Press, 1969).

Ginger, Ray. *Six Days or Forever? Tennessee vs. John Thomas Scopes*, Reprint edition (Chicago: Quadrangle Books, Quadrangle Paperbacks, 1969 [1958]).

Golding, G. "La Bible Contre Darwin: Le Procès Scopes aux Etats-Unis (1925)." *L'Histoire* 25 (June 1981): 18.

Grabiner, J. V., and P. D. Miller. "Effects of the Scopes Trial: Was It a Victory for Evolutionists?" 185 *Science* 4145 (6 September 1974): 832–837.

Grebstein, Sheldon Norman, ed. *Monkey Trial: The State of Tennessee vs. John Thomas Scopes* (Boston: Houghton Mifflin, 1960).

Hays, Arthur Garfield. "The Scopes Trial," in *Evolution and Religion*, Gail Kennedy, ed. (New York: Heath, 1957).

Holmes, S. J. "Proposed Laws Against the Teaching of Evolution." 13 *Bulletin of the American Association of University Professors* (December 1927): 549–554.

Loewenberg, Bert James. "Controversy Over Evolution in New England, 1859–1873." *New England Quarterly* 8 (June 1935): 232–257.

Rowell, Chester A. "The Cancer of Ignorance: The Spread of Anti-science in an American Commonwealth." IV *The Survey* 3 (1 November 1925): 159–161. (Touches on fundamentalist anti-evolution legislation, anti-vivisection legislation, and anti-medical legislation.)

Scopes, John T., and James Presley. *Center of the Storm: Memoirs of John T. Scopes* (New York: Holt, Rinehart, and Winston, 1967).

"Scopes Revisited." 109 *Commonweal* 4 (26 February 1982): 102, 104. (States that the theology of creationists does a "disservice" to the people it would represent.)

Tompkins, Jerry R., ed. *D-Day at Dayton: Reflections on the Scopes Trial* (Baton Rouge: Louisiana State University Press, 1965).

V. Creationism

A. Creationism as a Social Movement

Blinderman, Charles S. "Unnatural Selection: Creationism and Evolutionism." 24 *Journal of Church and State* 1 (1982): 73–86.

Cloud, Preston. "Evolution Theory and Creation Mythology." 37 *The Humanist* 6 (November/December 1977): 53–55.

"Creationism Evolves." 241 *Scientific American* 7 (July 1979): 72.

Godfrey, L. R. "Science and Evolution in the Public Eye." *The Skeptical Inquirer* (Fall 1971): 21–32.

Gurin, Joel. "The Creationism Revival." 21 *The Sciences* (April 1981).

"Is Darwin God?" 33 *National Review* 5 (20 March 1981): 265–266. (The Bible is no substitute for Darwin and vice-versa.)

Kitcher, Philip. *Abusing Science: The Case Against Creationism* (Cambridge, MA: MIT Press, 1982).

Mayer, William V. "The Emperor's New Clothes—Sold Again." 37 *The Humanist* 6 (November/December 1977): 52–53.

Michalsky, Walt. "The Masquerade of Fundamentalism." 41 *The Humanist* 4 (July/August 1981): 15–18, 52. (Makes the distinction that the Creationism controversy is not between religion and science but between Fundamentalism and science.)

Moose, T. M. "Creationist's Complaint Against the Movement." 99 *Christian Century* 5 (17 February 1982): 166–167. (A creationist disagrees with the movement's insistence on the incorporation of creation-science into public education, citing three contradictions of the theological assumptions at the heart of the matter.)

Olson, Edwin A. "Hidden Agenda Behind the Evolutionist/Creationist Debate." 26 *Christianity Today* 8 (23 April 1982): 26–30. (Creationism's "isolation" and "inflexibility" are doing more harm than good.)

Ruse, Michael. *Darwinism Defended: A Guide to the Evolution Controversies* (Reading, MA: Addison-Wesley, 1982).

Saladin, Kenneth S. "A Heretic in Dayton." 41 *The Humanist* 5 (September/ October 1981): 55–56. (Reflections on the "progress" made since the Scopes trial.)

Steinhart, Peter. "Fundamentals." 83 *Audubon* (September 1981): 5–14. (The creation-evolution dispute and the ethical and environmental consequences of ignoring evolutionary principles.)

Wolfe, Alan. "Creationism's Second Coming." 232 *The Nation* 11 (21 March 1981): 327–328.

B. Creationism in the Public Schools
Alexander, Richard D. "Evolution, Creation, and Biology Teaching." 40 *American Biology Teacher* 2 (February 1978): 91–96, 101–104.

Asimov, Isaac. "The Threat of Creationism." *New York Times Magazine* (14 June 1981): 90–101.

Bentley, Michael L. "On the Teaching of Origins." 70 *Today's Education* 2 (April/May 1981): 56G–58G. Also in the *Virginia Journal of Science*, the quarterly publication of the Virginia Academy of Science. (Recognizes an important point that creationists make: the theory of evolution, or any scientific theory for that matter, should not be taught as *fact*.)

Broad, W. J. "Creationists Limit Scope of Evolution Case." 211 *Science* 4488 (20 March 1981): 1331–1332.

Brush, Stephen G. "Creationism/Evolution: The Case AGAINST 'Equal Time.' " 48 *Science Teacher* 4 (April 1981): 29–33.

Brush, Stephen G. "Finding the Age of the Earth: By Physics or by Faith?" 30 *Journal of Geological Education* 1 (January 1982): 34–58.

"Californians Will Monitor Evolution Teaching in Schools." 26 *Christianity Today* 3 (5 February 1982): 69–70.

Cowen, Robert. "Creationism in the Classroom." *Technology Review* (July 1981): 8–9.

"Creationism Update." 40 *The Humanist* 4 (July/August 1980): 63–64. (Report on the annual meeting of the Georgia Academy of Sciences, April 1980.)

"The Creationists," special section in *Science 81* (December 1981), with articles on Creationism as a social movement ("The Genesis of Equal Time" by John Skow) and Creationism as science ("A Farewell to Newton, Einstein, Darwin . . ." by Allen Hammond and Lynn Margulis) plus sidebars on the impact on education and what the creationists say.

"Creationists Get Their Day in Court." 26 *Christianity Today* 6 (19 March 1982): 26. (Short discussion of why creationists believe that the teaching of evolution is a violation of their constitutional right to the free exercise of their religion.)

"The Creationist Threat: Science Finally Awakens." VI *The Skeptical Inquirer* 3 (Spring 1982): 2–5.

Dahlin, Robert. "A Tough Time for Textbooks." *Publishers Weekly* 220 (7 August 1981): 28–32.

Donohue, John W. "Creation and Evolution: 'A Balanced Treatment.' " 146 *America* 5 (6 February 1982): 90–92. (Creationists' strategy underlines failure of American public education to contribute to the religious literacy of students.)

Edwords, Frederick. "Creation/Evolution Update." 42 *The Humanist* 1 (January/February 1982): 46–47.

Edwords, Frederick. "Creation/Evolution Update: The Aftermath of Arkansas." 42 *The Humanist* 2 (March/April 1982): 55.

Eldredge, Niles. *The Monkey Business: A Scientist Looks at Creationism* (New York: Washington Square Press, 1982).

Gerlovich, J. A., *et al.* "Creationism in Iowa." (letter) 208 *Science* 4449 (13 June 1980): 1208–1211.

Godfrey, Laurie, ed. *Scientists Confront Creationism* (New York: Norton, 1982).

Jones, Bevel. "Science, Religion, and the Georgia Legislature." 98 *Christian Century* 1 (7–14 January 1981): 6–8.

Lewis, Ralph W. "Why Scientific Creationism Fails to Meet the Criteria of Science." *Creation/Evolution* (Summer 1981): 7–11.

Moore, John A. "Creationism in California." 103 *Daedalus* (Summer 1974): 173–190.

Moore, John A. "On Giving Equal Time to the Teaching of Evolution and Creation." 18 *Perspectives in Biology and Medicine* 3 (Spring 1975): 405–417. (Looks at 1973 Tennessee law.)

Moore, John N. "Evolution, Creation, and the Scientific Method." 35 *American Biology Teacher* (January 1973): 23–27.

Muller, Hermann J. "One Hundred Years Without Darwinism Are Enough." 19 *The Humanist* (1959): 139–155.

Nelkin, Dorothy. *The Creationism Controversy: Science or Scriptures in the Schools* (New York: Norton, 1982).

Nelkin, Dorothy. "Creation vs. Evolution: The Politics of Science and Education," in E. Mendelsohn, P. Weingart, and R. Whitley, eds., *The Social Production of Scientific Knowledge* (Dordrecht, Holland: Reidel, 1977).

Nelkin, Dorothy, "Science or Scripture: The Politics of 'Equal Time,' " in Gerald Holton and William Blanpied, eds., *Science and Its Public: The Changing Relationship* (Dordrecht, Holland: Reidel, 1976).

Nelkin, Dorothy. *Science Textbook Controversies and the Politics of Equal Time* (Cambridge, MA: MIT Press, 1977).

"Publishers, Lawyers Discuss Creationism in Textbooks." 219 *Publishers Weekly* 7 (13 February 1981): 26.

Saladin, Kenneth S. "Creationist Bill Dies in Georgia Legislature." 40 *The Humanist* 3 (May/June 1980): 59–60.

Saladin, Kenneth S. "Opposing Creationism: Scientists Organize." 42 *The Humanist* 2 (March/April 1982): 59, 61. (Chronicles the establishment of Committees of Correspondence.)

Sellers, James. "On Teaching Creation in the Schools." 98 *Christian Century* 42 (23 December 1981): 1333–1334.

Skogg, Gerald. "The Textbook Battle Over Creationism." 97 *Christian Century* 32 (15 October 1980): 974–976.

Wade, Nicholas. "Creationists and Evolutionists: Confrontation in California." 178 *Science* 4062 (17 November 1972): 724–729.

Zuidema, Henry P. "Genetics and Genesis: The New Biology Textbooks that Include Creationism." *Creation/Evolution* (Summer 1981): 18–22. (Followed by a discussion by biology educator Stanley Weinberg.)

C. Selected Creation-Science Publications, and Articles Espousing Creationism
Bird, Wendell. "Evolution in Public Schools and Creation in Students' Homes: What Creationists Can Do." *ICR Impact Series* 70 (April 1979); reprinted in Henry M. Morris and Donald Rohrer, eds., *The Decade of Creation* (San Diego: Creation-Life Publishers, 1981).

Bliss, Richard B. *Origins: Two Models—Evolution, Creation* (San Diego: Creation-Life Publishers, 1978).

Clark, M. E. *Our Amazing Circulatory System . . . By Chance or Creation?* (San Diego: Creation-Life Publishers, 1976).

Geisler, Normal L. "Creationism: A Case for Equal Time." 26 *Christianity Today* 6 (19 March 1982): 26–29.

Gish, Duane T. "Creation, Evolution, and the Historical Evidence." 35 *American Biology Teacher* 3 (March 1973): 132–140.

Gish, Duane T. *Evolution? The Fossils Say No!* (San Diego: Creation-Life Publishers, 1978).

Gish, Duane T. "The Scopes Trial in Reverse." 37 *The Humanist* 6 (November/December 1977): 50–51.

Hefley, J. C. *Are Textbooks Harming Your Children?* (Milford, MI: Mott Media, 1979).

Kofahl, Robert E., and Kelly L. Segraves. *The Creation Explanation: A Scientific Alternative to Evolution* (San Diego: Creation-Science Research Center, 1975).

Moore, J. *Should Evolution Be Taught?* (San Diego: Creation-Life Publishers, 1974).

Moore, John N., and Harold S. Slusher. *Biology: A Search for Order in Complexity* (Grand Rapids: Zondervan, 1970).

Morris, Henry M. *Evolution and the Modern Christian* (Philadelphia: Presbyterian and Reformed Publishing Company, 1967).

Morris, Henry M. *The Remarkable Birth of Planet Earth* (San Diego: Creation-Life Publishers, 1978).

Morris, Henry M. *Scientific Creationism* (San Diego: Creation-Life Publishers, 1974).

Morris, Henry M. *Studies in the Bible and Science* (Philadelphia: Presbyterian and Reformed Publishing Company, 1966).

Morris, Henry M. *The Troubled Waters of Evolution* (San Diego: Creation-Life Publishers, 1974).

Parker, Gary E. *Creation: The Facts of Life* (San Diego: Creation-Life Publishers, 1979).

Ramm, Bernard. *The Christian View of Science and Scripture* (Grand Rapids: Eerdmans, 1954).

Ramm, Bernard. "Theological Reactions to the Theory of Evolution." 15 *Journal of the American Scientific Affiliation* (1963): 71–77.

21 Scientists Who Believe in Creation (San Diego: Creation-Life Publishers, 1977).

Whitcomb, John C., and Henry M. Morris. *The Genesis Flood* (Philadelphia: Presbyterian and Reformed Publishing Company, 1961).

Wilder-Smith, A. E. *The Creation of Life* (San Diego: Creation-Life Publishers, 1970).

Wilder-Smith, A. E. *The Natural Sciences Know Nothing of Evolution* (San Diego: Creation-Life Publishers, 1981).

Willis, David L., ed. *Origins and Change: Selected Readings from the Journal of the American Scientific Affiliation* (Elgin, IL: American Scientific Affiliation, 1978).

Young, David A. *Christianity and the Age of the Earth* (Grand Rapids: Zondervan, 1982).

VI. Legal Perspectives on Anti-Evolution Laws

Gray, Virginia. "Anti-Evolution Sentiment and Behavior: The Case of Arkansas." 57 *Journal of American History* (September 1970): 352–366.

Le Clercq, Frederic S. "The Monkey Laws and the Public Schools: A Second Consumption?" 27 *Vanderbilt Law Review* 2 (March 1974): 209–242.

Sky, Theodore. "The Establishment Clause, Congress, and the Schools: An Historical Perspective." 52 *Virginia Law Review* 8 (December 1966): 1395–1466.

VII. The Arkansas Trial

Cook, Harry. "Monkey Trial Redivivus." 99 *Christian Century* 1 (6–13 January 1982): 6–7.

Cracraft, Joel. "Reflections on the Arkansas Creation Trial." 8 *Paleobiology* 2 (1982): 83–89.

"Creationists Lose in Arkansas." 26 *Christianity Today* 2 (22 January 1982): 28–29.

Eldredge, Niles. "Witnesses Weigh Textbooks at Arkansas Creationism Trial." 221 *Publishers Weekly* 1 (1 January 1982): 13–14.

Lewin, Roger. "A Response to Creationism Evolves," 214 *Science* 4521 (6 November 1981): 635–638. First of a series of reports by Lewin on the Creationism issue and the Arkansas trial. (See also the issues for December 4 and 11, 1981, and January 1, 8, 22, and 29, 1982).

Lyons, Gene. "Repealing the Enlightenment" 264 *Harper's* 1582 (April 1982): 38–40, 73–78.

Marsden, George M. "A Law to Limit the Options." 26 *Christianity Today* 6 (19 March 1982): 28–30.

Mutter, John. "Federal Judge Overturns Arkansas Creationism Law." 221 *Publishers Weekly* 4 (22 January 1982): 10.

Reuter, Madalyne. "ACLU Challenges Arkansas Act on 'Creationism' in Schools." 219 *Publishers Weekly* 24 (12 June 1981): 16.

Appendix A	Legislation Concerning the Teaching of "Creation-Science" and "Evolution-Science" in the Public Schools of Maryland

Excerpts from the Opinion of the Attorney General of the State of Maryland

[*Editor's note*. In 1981, legislation was proposed in the Maryland General Assembly (House Bill 1078) that would have required balanced classroom presentation of both "creation-science" and "evolution-science." In response to a request from legislator Patrick C. Scannello for a review of the bill's constitutionality, the Maryland Attorney General's Office issued a formal opinion on 23 February 1982, in a letter to Delegate Scannello.

The opinion, signed by Attorney General Stephen H. Sachs and Assistant Attorney General Ellen M. Heller, first reviews some of the provisions of House Bill 1078. The opinion is to be published at 67 *Opinions of the Attorney General of the State of Maryland* (1982). The editor thanks the Office of the Attorney General, and especially Ellen M. Heller, Esq., for help in reprinting these excerpts.]

You have asked us to review for constitutionality House Bill 1078 concerning the teaching of "creation-science" and "evolution-science" in the public schools of Maryland. In particular, you have asked whether this bill violates the First Amendment of the United States Constitution, which requires that "Congress shall make no law respecting an establishment of religion or prohibiting the free exercise thereof."

The bill has a number of components, but its key feature provides that public schools are required to give a "reasonably unbiased" presentation of "creation-science" and "evolution-science" in classroom lectures, textbook materials, and library volumes. . . . "Reasonably unbiased presentation" in classroom lectures is defined in pertinent part as the presentation of "evolution-science" and "creation-science" by the "time or length balance provided by an equal number of class periods or course hours to the nearest hour in a school year". . . . For textbook materials, "reasonably unbiased" treatment requires the same "content balance" as in classroom lectures but, additionally, mandates "the length balance provided by an equal number of pages . . . to the nearest 10 percent of the total pages in assigned textbook materials," to the extent such materials are available. . . . The "content balance" required in library volumes is that "provided by an equal number of library volumes" in the humanities and sciences presenting evolution and creation science, to the extent such volumes are available.

For the reasons discussed below, we have concluded that House Bill 1078 violates the separation between Church and State required by the "Establishment Clause" of the First Amendment.[1][*Editor's note.* This excerpt retains the original footnotes and footnote numbers from the full Opinion of the Attorney General.]

The Supreme Court and the Establishment Clause

The Supreme Court has stated that the "first and most immediate purpose [of the Establishment Clause] rest[s] on the belief that a union of government and religion tends to destroy government and to down-grade religion. . . . Another purpose of the Establishment Clause rest[s] upon an awareness of the historical fact that governmentally-established religions and religious persecutions go hand in hand" (*Engel v. Vitale* at 431). The Court has given a broad meaning to the Establishment Clause and has held that it means, at the very least, that "neither the States nor the Federal government can pass laws which aid one religion, aid all religions, or prefer one religion over another" (*Everson v. Board of Education* at 15–16). Its effect has been "to take every form of propagation of religion out of the realm of things which could directly or indirectly be made public business" by comprehensively forbidding all forms of public support for religion (*Idem* at 26, 31–32).

Because the public schools have been characterized "as a most vital civic institution for the preservation of a democratic system of government," the Establishment Clause has encountered its severest tests

---

1. We have recently discussed at some length the meaning and scope of the 'religious clauses' of the First Amendment. See 65 *Opinions of the Attorney General* 186 (1980) (The distribution of Gideon Bibles to public elementary school students violates the Establishment Clause); 64 *Opinions of the Attorney General* 134 (The teaching of ethical values in public schools is constitutionally permissible).

when it has been applied to practices and policies within the classroom (*School District of Abbington Township, Pennsylvania v. Schempp* at 230). . . .

Separation of religion in the field of education was not imposed upon unwilling states but was the democratic response of the American community to the particular needs of harmonizing "multiform creeds." "Designed to serve as perhaps the most powerful agency for promoting cohesion among the heterogenous democratic people, the public schools must keep scrupulously free from entanglement in the strife of sects" (*McCollum v. Board of Education* at 216–217).

The Supreme Court has enunciated a three-part test for determining whether a challenged State practice or policy is permissible under the Establishment Clause:

"First, the statute must have a secular purpose; second, its principal or primary effect must be one that neither advances nor inhibits religion. . . ; finally, the statute must not foster 'an excessive government entanglement with religion' " (*Lemon v. Kurtzman* at 612–613).

If a statute violates any one of these three principles, it must be struck down under the Establishment Clause (*Stone v. Graham*). . . . [*Editor's note.* The Maryland Opinion then reviews the U.S. Supreme Court's ruling in *Epperson v. Arkansas* (1968), which challenged a previous Arkansas "anti-evolution" statute, and outlines the case of *McLean v. Arkansas* (1982), quoting extensively from Judge Overton's opinion.]

The trial before the United States District Court for the Eastern District of Arkansas . . . serves as a valuable laboratory in which scientific creationism, as a mandated subject of public school instruction, has recently been tested against the Establishment Clause of the First Amendment.

Purpose

Judge Overton found that the Arkansas statute "was simply and purely an effort to introduce the Biblical version of creation into the public school curricula" and thus "passed with specific purpose by the General Assembly of advancing religion" . . . . An important element he considered was the motivation of the bill's actual author, the head of a scientific creationist organization called "Citizens for Fairness in Education."[5] The evidence adduced at the Arkansas trial revealed that the author's efforts in preparing model acts providing for the teaching of creationism and crusading for their adoption were motivated by his opposition to the theory of evolution and his desire to see the Biblical version of creationism taught in the public schools. . . .

House Bill 1078 is substantially based upon—and, in several aspects, virtually identical with—the Arkansas statute invalidated in *McLean*. Despite certain changes in style, tone, and language, both share the same ultimate purpose. This is apparent whether one considers the source of the bill, sponsor statements in published reports [See, for example, "Teachers fight bill on creationism," *Baltimore News-American*, 16 October 1981, p. 1A] or, most importantly, the language contained in the bill's preamble. For example, the preamble to House Bill 1078 emphasizes that evolution-science is contrary to the religious beliefs of many students and parents. . . ; that failing to teach creation-science abridges freedom of religious exercise and undermines religious beliefs . . . ; and that the teaching of only evolution-science discriminates against religions that profess creationist beliefs. . . . It is impossible for us to read such statements without perceiving a religious purpose behind the bill.

It is true that a stated purpose in the preamble is to permit a "reasonably unbiased presentation of creation-science" and evolution; "to ensure academic freedom . . . to make public schools neutral toward students' diverse philosophical beliefs . . . to end any establishment of religion. . . ." But such lofty disclaimers cannot wish away what to us is the plain fact that religious conviction is the preeminent purpose of House Bill 1078, just as it was found to be the purpose of the Arkansas statute. "[A]n 'avowed' secular purpose is not sufficient to avoid conflict with the First Amendment" (*Stone v. Graham*).

5. We have learned that the author of the Arkansas bill, Paul Ellwanger, supplied the model bill that was the basis for House Bill 1078. As mentioned in the *McLean* opinion, Mr. Ellwanger began in 1977 to collect and propose legislative 'model state acts' requiring the teaching of scientific creationism. He believed that evolution is the 'forerunner of many social ills, including Nazism, racism, and abortion' and that the crusade to teach creationism in the schools is a 'battle . . . between God and anti-God forces'. . . .

Effect

House Bill 1078 also fails under the second prong of the Establishment Clause test in that its language leaves no doubt that the effect of the bill would be to promote the advancement of religion in public schools.

First, like the definition of "creation-science" in the Arkansas statute invalidated by Judge Overton, the definition of "creation-science" in House Bill 1078 unequivocally reflects the influence of the Book of Genesis. The Arkansas bill described creation-science as "sudden creation of the universe, energy and life from nothing," and, as the Court concluded, the statement could only imply a supernatural creation by God. Although House Bill 1078 does not use the words "life from nothing," it still necessarily conveys the religious concept of *creatio ex nihilo*, or sudden creation from nothing. . . .

Second, House Bill 1078, consistent with the general position of creationists, takes the approach that there are only two explanations for life as it presently exists—the theory of evolution or the Biblical creationist belief—and that both of these theories can be taught in a scientific manner. The bill's preamble states:

"WHEREAS, Creation-science is an alternative to evolution-science as an explanation for origins, and *creation-science like evolution-science can be presented from a strictly scientific standpoint without any religious doctrine* because there are qualified scientists who conclude that scientific evidence better supports creation-science just as there are qualified scientists who reach similar conclusions about evolution-science, and because scientific evidence and related inferences have been presented by qualified scientists for creation-science just as they have been presented for evolution-science . . ." (emphasis supplied) . . .

However, the evidence at the Arkansas trial simply does not support the proposition that creation-science *is* a scientific theory or *can* be presented from a scientific viewpoint. Rather, it seems unequivocally a religious doctrine. The emphasis on "origins" as a theory of evolution is peculiar to creationist literature and the scientific community does not consider the actual origins of life as part of the evolutionary theory. "The theory of evolution assumes the existence of life and is concerned with various explanations of *how* life evolved." . . .

As the *McLean* opinion points out, creation-science, with its emphasis on creation rather than evolution, is not a scientific theory "because it depends upon a supernatural intervention which is not guided by natural law. It is not explanatory by reference to natural law, is not testable, and is not falsifiable"... . No scientific journal has ever published an article espousing the creation-science theory. Indeed, House Bill 1078 states that "creation-science is not an unquestionable fact of science" and acknowledges that it cannot be "experimentally observed, fully verified, or logically falsified." It makes the same assertions concerning the theory of evolution, but they are rhetorical statements *without* the support of the scientific community.

Third, the bill does not avoid constitutional infirmity by substituting "reasonably unbiased" presentation for the Arkansas requirement of "balanced" treatment. Reasonably unbiased presentation, as defined, clearly provides that "equal treatment" will be given creation-science and evolution. However, "[s]eparation is a requirement to abstain from fusing functions of Government and of religious sects, not merely to treat them all equally" (*McCollum v. Board of Education* at 227).

Thus, the lack of scientific authority and the inherent religious nature of "creation-science" has led us to conclude that requiring its teaching in the public schools would not only have the effect of advancing religion but would promote one religious belief over another.

Entanglement

Finally, we believe that the implementation of the bill would cause serious problems under the entanglement prong of the Establishment Clause test. The bill requires "content balance" and "reasonably unbiased presentation" of the theory of evolution and creation-science. As defined, this means equal time, equal number of class periods or hours, equal number of pages in texts, and equal number of library volumes dealing with both these theories. In regard to the text and library volumes, the bill purports to require equality only to the extent such volumes are available. However, there may be great difficulties in finding materials on creation-science that would be appropriate for school use. In Arkansas, the individual assigned the duty of preparing a curriculum guide on creationism found that all of the available materials were unacceptable because "they were permeated with religious references"... .

Such legislation involves an enormous potential for government entanglement with religion. Supervisors will have difficulties ensuring that teachers devote exactly the same number of hours to the teaching of evolution-science and creation-science... .

One might also ask: How could we justify mandating the teaching of "creation-science" in our public schools and not, for example, also mandate the counterpart teachings of traditions such as Hinduism, Confucianism, and Islam? And, if those teachings are to be taught, what of the more controversial teachings of the Reverend Sun Myung Moon and the Unification Church, or the teachings of the Science of Creative Intelligence — Transcendental Meditation . . .?

All of this is the essence of religious entanglement. And that is precisely what is forbidden by the First Amendment. As we have previously noted:

"The strain placed on school authorities by requiring them to make such religious judgments and to monitor their enforcement would entangle them in conflicts among religious groups, with the schoolhouse as the field of battle" [65 *Opinions of the Attorney General* 186, 196 (1980)].

Conclusion

For all these reasons, it is our view that House Bill 1078 has as its purpose and effect the advancement of religion and would foster an excessive governmental entanglement with religion.

Because we are "a religious people" (*Zorach v. Clauson* at 313), the notion of government neutrality toward religion is often difficult to accept. But we must remember that many of our forebears came to these shores to escape persecution at the hands of hostile governments and that we are today a nation of diverse, frequently conflicting, religious faiths. Then, perhaps, we can better understand that the bond that unites us all is our agreement that government should be neutral in matters of religion. The price of religious liberty, in short, is official neutrality.

Appendix B   Act 685 of 1981

                              Louisiana Revised Statutes

[*Editor's note*. On 10 July 1981 the Governor of the State of Louisiana signed into law an amendment to the state's General School Law. The declared legislative purpose of the act was to require "balanced treatment of creation-science and evolution-science in public schools," "to bar discrimination on the basis of creationist or evolutionist belief," and to provide "inservice teacher training and materials acquisition" relative to curriculum development on these subjects. The law is currently under court challenge at the state level.]

Be it enacted by the Legislature of Louisiana:

Section 1. Subpart D-2 of Part III of Chapter I of Title 17 of the Louisiana Revised Statutes of 1950, comprised of Sections 286.1 through 286.7, both inclusive, is hereby enacted to read as follows:

CHAPTER I. GENERAL SCHOOL LAW

PART III. PUBLIC SCHOOLS AND SCHOOL CHILDREN

SUBPART D-2. BALANCED TREATMENT FOR CREATION-SCIENCE AND EVOLUTION-SCIENCE IN PUBLIC SCHOOL INSTRUCTION

§286.1. Short Title
This Subpart shall be known as the "Balanced Treatment for Creation-Science and Evolution-Science Act."

§286.2. Purpose
This Subpart is enacted for the purposes of protecting academic freedom.

§286.3. Definitions

As used in this Subpart, unless otherwise clearly indicated, these terms have the following meanings:

1. "Balanced treatment" means providing whatever information and instruction in both creation and evolution models the classroom teacher determines is necessary and appropriate to provide insight into both theories in view of the textbooks and other instructional materials available for use in his classroom.

2. "Creation-science" means the scientific evidences for creation and inferences from those scientific evidences.

3. "Evolution-science" means the scientific evidences for evolution and inferences from those scientific evidences.

4. "Public schools" mean public secondary and elementary schools.

§286.4. Authorization for balanced treatment; requirement for nondiscrimination

A. Commencing with the 1982–1983 school year, public schools within this state shall give balanced treatment to creation-science and to evolution-science. Balanced treatment of these two models shall be given in classroom lectures taken as a whole for each course, in textbook materials taken as a whole for each course, in library materials taken as a whole for the sciences and taken as a whole for the humanities, and in other educational programs in public schools, to the extent that such lectures, textbooks, library materials, or educational programs deal in any way with the subject of the origin of man, life, the earth, or the universe. When creation or evolution is taught, each shall be taught as a theory, rather than as proven scientific fact.

B. Public schools within this state and their personnel shall not discriminate by reducing a grade of a student or by singling out and publicly criticizing any student who demonstrates a satisfactory understanding of both evolution-science or creation-science and who accepts or rejects either model in whole or part.

C. No teacher in public elementary or secondary school or instructor in any state-supported university in Louisiana, who chooses to be a creation-scientist or to teach scientific data which points to creationism shall, for that reason, be discriminated against in any way by any school board, college board, or administrator.

§286.5 Clarifications

This Subpart does not require any instruction in the subject of origins but simply permits instruction in both scientific models (of evolution-science and creation-science) if public schools choose to teach either. This Subpart does not require each individual textbook or library book to give balanced treatment to the models of evolution-science and creation-science; it does not require any school books to be discarded. This Subpart does not require each individual classroom lecture in a course to give such balanced treatment but simply permits the lectures as a whole to give balanced treatment; it permits some lectures to present evolution-science and other lectures to present creation-science.

§286.6 Funding of inservice training and materials acquisition

Any public school that elects to present any model of origins shall use existing teacher inservice training funds to prepare teachers of public school courses presenting any model or origins to give balanced treatment to the creation-science model and the evolution-science model. Existing library acquisition funds shall be used to purchase nonreligious library books as are necessary to give balanced treatment to the creation-science model and the evolution-science model.

§286.7. Curriculum Development

A. Each city and parish school board shall develop and provide to each public school classroom teacher in the system a curriculum guide on presentation of creation-science.

B. The governor shall designate seven creation-scientists who shall provide resource services in the development of curriculum guides to any city or parish school board upon request. Each such creation-scientist shall be designated from among the full-time faculty members teaching in any college and university in Louisiana. These creation-scientists shall serve at the pleasure of the governor and without compensation.

Section 2. If any provision or item of this Act or the application thereof is held invalid, such invalidity shall not affect other provisions, items, or applications of this Act which can be given effect without the invalid provisions, items, or applications, and to this end the provisions of this Act are hereby declared severable.

Section 3. All laws or parts of laws in conflict herewith are hereby repealed.

Index of Names and Organizations

Index of Legal Cases